HARVEY C. FRIEDENTAG

STOCKS for
OPTIONS
TRADING

Low-Risk, Low-Stress Strategies for Selling Stock Options — **PROFITABLY**

S^{t}_{L}

St. Lucie Press
Boca Raton   London   New York   Washington, D.C.

AMACOM

American Management Association
New York • Atlanta • Boston • Kansas City • San Francisco • Washington, D.C.
Brussels • Mexico City • Tokyo • Toronto

Library of Congress Cataloging-in-Publication Data

Friedentag, Harvey, C. (Harvey Conrad)
  Stocks for options trading : low -risk, low-stress strategies for selling stock options
profitably / Harvey C. Friedentag.
    p.cm.
  Includes bibliographical references and index.
  ISBN 0-910944-07-5 (CRC Press LLC : alk. paper) —
  ISBN 0-8144-0423-5 (Amacom : alk. paper)
  1. Stock options. I. Title.
HG6042.F74 2000
332.63´228—dc21   99-048612
                                                                                  CIP

© 2000 by CRC Press LLC
St Lucie is an imprint of CRC Press LLC

No claim to original U.S. Government works
International Standard Book Number 0-910944-07-5 (CRC Press LLC)
International Standard Book Number 0-8144-0423-5 (Amacom)
Printed in the United States of America 1 2 3 4 5 6 7 8 9 0
Printed on acid-free paper

PREFACE

The greatest investment experience of your life begins today. It is really something timely, because all the "day-traders" are gambling without knowing what they are doing.

Twenty-four years ago, working at my kitchen table I started a unique endeavor for investing. Since that modest beginning, my portfolio management service has grown to become a source of personal satisfaction and income not just for me but, even better, for my clients.

Why? Because it covers what I know to be the single most rewarding form of investing in existence—the covered call option.

Let me open your eyes to the facts about this extraordinary investment arena. In the following pages you will learn the basics of selecting the best stocks for writing call options.

This book is a map. I have developed a step-by-step program to make you a successful investor, as I have made my clients successful investors. Let me lead you to the same success.

The approach is simple. The complete strategy is easy to execute and will save you both time and money. Section I of this book is truly fundamental, dealing with the basics of investing. If you're already comfortable with investing, I suggest you move directly to Section II, which tells you how to move beyond comfort to real success.

The methods discussed in this book have achieved proven results over a number of years. The strategies and tactics outlined will allow you to accumulate assets steadily. You can reach your investment goals within a reasonable amount of time. In studying my book you will learn:

- from the past,
- to invest without doubt,

- to make profits in the stock market, and
- to profit whether the market goes up, down, or sideways.

If that's what you're looking for, read on.

Harvey Conrad Friedentag

About the Author

The author is a Registered Investment Advisor (RIA) with the United States Securities and Exchange Commission and has been managing personal portfolios professionally since 1986. Mr. Friedentag has been certified as a Federal Court expert witness on stock trading; is an acknowledged expert on stock renting, the use of derivatives (exchanged-traded equity call options); and serves a President of the Contrarian Investment Club in Denver, Colorado. He is also author of *Investing Without Fear: Options* (International Publishing Corp., Chicago).

ACKNOWLEDGMENTS

A writer writes, a reader reads. As a writer, you know who you are and what the material means to you. But you cannot begin to know who the reader will be, or what background that reader will or won't bring to the page.

I have tried to write a book that will be useful and valuable to both current and future investors. However, this book would not have been possible without the assistance of many individuals.

There are not enough words in the thesaurus to thank Hal and Martha Quiat for their help in desktop publishing, grammar, and making me clearly state what I meant to say. They were able to divorce their own egos and personalities from the process and concentrate on drawing me out, which they did flawlessly. Bold and merciless editing to hone the material, to be sure that only the best parts of the document were included, was matched with a stubborn determination to reject anything that did not belong. Behind their engaging and gentle exteriors lies a strict discipline to prohibit excess wordage. Hal and Martha's weeding and pruning have been crucial to the presentation of this material. They gave freely of their time and talent rearranging their workday to accommodate mine.

I also have to thank some long time friends: Stuart MacPhail for getting me computerized, editing my first drafts, correcting my grammar, and helping me get my thoughts organized; Nate Oderberg for being there and reviewing my strategies; Thomas Tolen for helping me with the tax ramifications of investing; and Marty Shure, who bounced strategies and ideas off my head and really enhanced my investment procedures.

Finally, I want to thank my wife Beverly and my family for all the long sessions when I disappeared into my office to work on this book and was not available to them.

Table of Contents

SECTION I
THE BASICS

CHAPTER 1
TIME AND MONEY

Put not your trust in money, but put your money in trust.
O. W. Holmes, Jr.

Right after a day was discovered, someone subdivided it into two parts. Rather than leave it as simple as night and day, someone else sold us on measuring it into three phases: work, play, and sleep.

The work period is designed to force us to produce enough income to provide us with money for play and a place to sleep. The play period gives us time to unwind from the pressures of work. The purpose of the play period is to make us obsessed with having fun by spending all the money we earn at work. Most of us view these two periods as the useful phases of life. Sleep isn't considered to be a productive activity.

We tend to ignore or hide the fact that we sleep at all. We want to work long enough to earn enough money to go out and play. When the money is gone, we can always go back to work and earn enough to get back to play.

Working and playing hard are admirable. Does anyone work or play hard for the reward of sleep? If it were really popular, would we leave it as the last thing to do each day? Sleep seems to be something to do when there's nothing else to do.

The point is this: You manage to combine work and a little play. When you're sleeping, you aren't able to play or work. Nearly everyone has the ability to combine work and play, but very few know how to get anything but rest out of sleep.

Sleep robs us of time we could spend or want to spend, working or playing. Is sleep a waste of time? If we could find a better way to rest, probably we would. Since we can't, why not find something that will yield rewards for us while we're sleeping?

There is a way to make money while we sleep so that the entire day is productive. You'd never dream just how easy and uninvolved it is.

It's a method of letting something else work for you all the time, even when you're asleep. It involves very little work on your part. It involves making money, which is enjoyable, and it's one of the few possible methods of combining sleep and play.

That's what this book is all about. It's a how-to book. More than that, it's a how-to-know-how-to book. When you've finished this book, you'll have proof of a method of investing that you can either dream about or do something about. You will learn the difference between the haves and the have-nots—which boils down to the dids and the did-nots.

BECOME A DOER

Money makes money, and the money the money makes, makes more money. Benjamin Franklin

This book is intended to teach you about investing and how my option strategy can work for you. It's fantastic, unbelievable, excellent, super, fun, exciting, and most of all profitable.

Yes, you can invest your money in real estate and certificates of deposit. However, those who have benefited most own stocks. Some of those who benefited own shares of stocks through their pension funds; others have been in the market since the mid-1980s.

The stock market has quadrupled since 1983. There are few general investment classifications that offer a wide variety and range of possibilities in the stock market. The stock market offers opportunities for investment with every degree of risk, yield, growth, predictability, liquidity and tax shelter. Regardless of what your goals are, there is a part of American business—shares of stock—that can be acquired to let you reach your goals. Everyone interested in financial success can find it in the stock market.

In order to understand the investment purposes of the stock market, you need to understand the underlying purpose of the stock market. The stock market has the basic financial purpose of building homes, growing and processing the food we eat, building businesses, factories, and machines, and producing and delivering energy. We need the stock market to finance the building of recreational vehicles, road equipment, TVs, TV networks, stereos, computers, clothing, and all the support for all the fun things we wish to do—everything we need from the womb to the tomb.

I'm confident generalizing that there's an opportunity in the market for almost every type of investor. It needs only minimal management and liquidity can be excellent. Long-term growth is wonderful. In addition, there's predictability to an extremely high degree. Risk can be low and easily controllable. Yield is as a rule higher than any other equivalent investments. With it, tax shelters can be quite attractive. The secret is my option strategy.

In brief, it can be said that my strategy offers the greatest appeal to investors because it offers low risk, great predictability, moderate to high yield, excellent liquidity, substantial tax shelters, and high growth. In addition, management of the investment ranges from moderately passive to highly active.

The strategy is simple to use, can provide income in flat markets, income and capital gains in rising markets, and some safety of capital in falling markets.

No matter in which direction stocks move, you stay in positive territory. This method can provide double-digit returns regularly.

UNDERSTANDING COVERED CALLS

This is a method that routinely uses covered calls. This method of investing is used to capture price movements to produce steady results.

It begins by assembling a portfolio of good, large company stocks that can be used to support covered calls (options to sell stocks at a fixed price and for a specified time). Covered call options, like stocks, trade constantly on the option exchanges for periods of up to three years; though some consider them risky, covered calls can be used to execute very conservative strategies. Just don't buy them!

Covered calls work this way: Say you own ABC Corp., which is priced at $34.50. You can sell someone a covered call that gives the buyer the right to buy the stock at $40 any time during the next six months. For that privilege, the buyer pays $3.25 a share up front. If the stock goes nowhere, you, the seller, keep the shares—and the income from the covered call. If the stock rises to $50, the buyer pays only $40 for the shares and pockets a big profit, but you don't do badly. . . . By selling covered calls you're guaranteed income though you may give up the ability to benefit from big market rises.

The rise in the stock, $34.50 up to $40, produces capital gains in the profit from the up-front payment for the covered calls and the selling of the stock at the higher price. That income stream swells when the price of calls goes up. What is important is not whether the stock goes up or down, but how volatile it becomes. When stocks are volatile, the price of covered calls skyrockets. This is because option buyers, betting there's a good chance the stock will record the kind of jumps that make call options produce windfalls, are willing to pay more for them. When stocks stay even, the price of call options lags and you earn less.

If, after you buy a stock, you want to hold it but are worried that the price could go down, you can sell covered calls below the market price to generate income immediately. This protects you against a price downturn and if you are correct the below-market price of the covered call works like insurance, since you received more money up front when you sold the covered call.

Earnings from sales of covered calls have been as low as 9 percent when stocks have been motionless and as high as 40 percent when stocks have gone wild.

A STEADY PERFORMER

My uncommon approach dates to 1973 when the first exchange devoting itself exclusively to options trading—the Chicago Board Options Exchange (CBOE)—came into being. At the CBOE options could be traded on the floor just as stocks are traded on the stock exchanges. Option investors now enjoyed exchange and regulatory protection like that afforded stock investors. Most important, options could be bought and sold on the exchange at any time. Thus, a covered-call options seller could start taking in cash premiums from the eager options buyers.

Some investors use covered calls as an alternative to bonds. That makes sense because they are less volatile than most intermediate-term bonds—covered calls aren't sensitive to rate movements. Using covered calls will keep you in the black during periods of rising rates and my covered call strategy will be an unusually steady performer in all kinds of markets. As long as covered call prices remain in their normal range this strategy should continue to produce reliable double-digit results in good times and bad. Use covered calls to reduce risk and protect assets, not to aim for eye-catching returns (though they sometimes happen). Personally, I have never deviated from this.

It's fantastic. It's unbelievable. It's excellent, super, fun, exciting—and most of all, profitable.

THE COVERED CALL STRATEGY

When long- and short-term interest rates are low and yields on common stocks even lower, what should yield-hungry investors do? Buy stock, then sell covered calls on a share-for-share basis.

Covered call writing is appropriate for income-oriented and nervous investors. With covered call writing, calls may be sold against stock already or soon to be owned.

How does the strategy work?

Assume that you buy stock ABC at $9. It pays a 2 percent dividend, about 5 cents per quarter. You then sell the 90-day $10 covered call ($10 is the strike price or selling price) for $1.

If the stock remains unchanged after 90 days, the covered call writer—you—will receive 5 cents in dividends and retain the $1 premium someone paid up front for the covered call. This is a total of $1.05 per share (not including transaction fees). Over a 90-day period this is a static return of about 11.5 percent, or a 46 percent annualized return if the same rate continued for a year.

What if the stock rises above $10? You still get the call premium of $1 and the stock price appreciation of $1 per share, the difference between your purchase price of $9 and the $10 strike price of the call. This brings your total profit to $2.05 per share (not including transaction fees) or about 23 percent in 90 days. If you can do a similar transaction every 90 days over the course of a year, this equates to an annualized return of 92 percent.

In Summary

Buy ABC stock for $9.
Sell a covered call on your ABC stock for a $1 cash premium.
If called, sell ABC stock for $10.
$1 appreciation in stock + $1 cash premium = $2 profit.
$2 profit + .05 dividend = $2.05.

There are a few risks: If the stock declines, the cash premium received from selling the covered call offers limited protection. But it does lower the investor's breakeven price to $8 per share.

ABC stock cost = $9 − $1 premium = $8 out of pocket

Below that price, the position is nearly the same as owning the stock without a call, and the covered call investor must decide on an appropriate course of action: hold the position, buy back the covered call and write another call at a lower strike price, or close out the position altogether.

A covered call investor must be willing to own the stock. This means the investor is confident about the stock's prospects but can also accept being wrong. The income-oriented covered call investor who is suited for such risks could withdraw some or all the cash premium, after transaction fees, with reserves maintained for taxes.

Investing is like riding a bicycle. You don't fall off unless you plan to stop pedaling. Claude Pepper

This is not a buy-and-hold strategy. It involves thinking ahead about the appropriate action to be taken when the call period expires or if a stock price declines. If the covered call expires, an equal amount of thought is required to decide whether to write another covered call or unload the stock and invest elsewhere, either in another covered call or in another stock. With covered call writing, the goal is income, so the risk of stock ownership must be managed very carefully. Investors should focus on quality stocks.

The next step is to find covered call cash premiums that can be sold at acceptable rates of return.

All the details for handling each step will be explained thoroughly in Section II, from stock market fundamentals, to finding the right stock to own, to writing a covered call.

You can't just rush headlong into investing. You need to use a magnifying glass and a scale so you can examine the investment thoroughly before you act. As an investor, you should also learn what history has taught. In the next chapter, I'll briefly review these lessons.

A man that is young in years may be old in hours, if he has lost no time.
Francis Bacon

There's nothing wrong with cash, it gives you time to think.
Robert Prechter

CHAPTER 2
FEAR AND
IGNORANCE

Fear always springs from ignorance. Ralph Waldo Emerson

Today a growing number of people are entranced with the subject of investing—without having any basic know-how. Most become amateur investors; many find themselves caught in a poor deal, afraid to invest again; few become successful.

Making the leap from serious to successful investor, while seemingly a simple and evolutionary step, is quite complex. The desire to become a successful investor can be strong, but with it comes the task of developing an investment strategy. It is *how* you pursue your interest that usually sparks the jump from casual investor to successful status. Even if you are not going to become a professional investor, you have to think and invest like one!

No matter what the strategy, stock selection demands fundamental knowledge. What you see today in the economy of our country and the stock market has happened before. It will happen again. You can learn the fundamentals by understanding some basic stock market economic conditions and their meanings.

Stock cycle is a period of expansion (recovery) and contraction (recession) in economic activity, which affects inflation, growth, employment, and the prices of stocks.

Bull market is a period of prolonged rise in the prices of stocks, bonds, or commodities.

Bear market is a prolonged period of falling prices. A bear market in stocks is usually brought on by the anticipation of declining economic activity.

A stock market crash, the ultimate in bear markets, results from a complex interplay of the mass psychology of investors and the underlying fundamentals of the market. Pre-crash periods show common patterns of excessive optimism among investors, widespread use of borrowed money to buy stocks, and gradual but largely unrecognized erosions of the foundations of the market. Exactly when the market will fall can't be predicted, but the odds of a collapse can be estimated.

What I find most interesting today is the absence of memory of "Black Monday," October 19, 1987. As a financial professional, it was one of the most exhilarating times in my career. It was the day that the financial world stumbled and forced a new understanding of financial markets and their global impact. It shook those stubborn stock traders from their 1980s highs and expanded their view beyond Wall Street.

Many of today's investors who weren't around during the 1987 crash have basked in the sunny U.S. economy with few worries, but I wager that those who survived through 1987 keep looking over their shoulders. I say that there is too much money in this market, but when it crumbles, it will fall in sectors, not all at once as it did in 1987. Today price/earnings ratios are at historically high levels. Historical volatility is also at levels as high as those levels just before the 1987 crash. As one trader said to me after the 1987 crash, "You have all these signals that everybody traditionally looked at, and investors were lulled into this idea that liquidity will drive the market, so damn the P/Es—full speed ahead."

Granted, Black Monday didn't end the bull market, but it did change many lives. So the question is, if it isn't the end of the bull run, is another 22 percent break around the corner?

LESSONS FROM HISTORY

The best opening lesson is illustrated by the largest money-losing event and the worst day of panic ever seen—The Stock Market Crash of 1929.

The 1929 Market Crash

The most dramatic market setback, the 1929-1932 Crash, saw the Dow Jones Industrial Average drop nearly 90 percent. Here's what happened:

Tuesday, October 29, 1929: The average dropped $10, to $70. Shortly after 1 p.m., prices were down 1 to 50 points on both the New York Stock Exchange and the Curb Exchange. During the day slight recoveries set in, but these were without support and the stocks that had recovered fell back. By 2:10 p.m. 13,838,000 shares had changed hands. At this time, the stock ticker tape was 82 minutes late. There was a mid-afternoon rally from the lows, which brought prices back from the minimums but still left them down enormously on the day.

Wednesday, October 30, 1929: The average advanced $5, to $30. Investment trusts and trading corporations were heavy buyers of stock on both Tuesday and Wednesday. Estimates of total purchases ranged from $350 million to half a billion dollars. On Wednesday, the bulls staged a greed demonstration in the closing minutes. Prices were whirled up to the highs of the day, a day that had seen prices moving up regularly from their lows of Tuesday.

The nation's leading financial forces mobilized to calm the wave of hysteria and restore confidence in the securities markets. John D. Rockefeller, Sr. and his son announced that, for some days, they had been purchasing sound common stocks. Julius Rosenwald, philanthropist and chairman of the board of Sears, Roebuck and Company, "pledged, without limit" his personal fortune to guarantee the stock market accounts of the 40,000 employees of his company (a plan he had also adopted during the depression in 1921).

Thursday, October 31, 1929: The three-hour stock exchange session on Thursday saw traders push the market forward at such a pace that $10 billion was added to the market's valuation of stock, and profit-taking failed to wipe out the gains. The first half-hour alone was at a rate of more than 24 million shares. Tickers ran an hour behind, but floor quotations at closing time showed stocks were up from 1 to 40 points. Buying was as frenzied on Thursday as selling had been on Tuesday. Values came back with the vigor of the old bull market that Wall Street had declared dead a few days before.

The following comparison illustrates that despite the losses a lot of money could have been made as a result of the 1929-1932 Market Crash:

	Dow Jones	Percent Change
October 29, 1929	−30.57	−11.73
October 30, 1929	+28.40	+12.34

Financial hell or financial heaven may be just around the corner; however, a prepared investor can profit when the market goes up *or* down. It's best to emulate investors like John D. Rockefeller, Sr. and Julius Rosenwald.

Market Madness: Black Monday, October 19, 1987

More than 58 years later, the 1987 stock market crash sent the Dow Jones Industrial Average (DJIA) plummeting 508 points, to 1,738.74; total market losses climbed to more than half a trillion dollars. Over the next two days the market came back slightly but headed back down on Thursday,

October 22, 1987 when in heavy trading the Dow closed down 77.42 points, at 1,950.43.

Among the issues most heavily traded was IBM. On Tuesday, trading in the stock was halted to stop the flow of panic selling. Several other companies reacted to the debacle by buying back large portions of their shares at discounted prices in order to show confidence in their future— and discourage takeovers.

If you had bought IBM on Monday, October 19, you would have profited in just two days. The following table illustrates your net gain if you have taken positions in three companies:

Company	Oct. 19	Oct. 21	Net Gain
IBM	103 $^1/_4$	122 $^3/_4$	19.50
AT&T	23 $^5/_8$	29 $^1/_2$	5.87
Bell South	33 $^3/_4$	39 $^1/_2$	5.75

While almost no company's stock escaped unscathed, the utilities did not get hit as hard as other sectors. Utilities are considered to be safe, or defensive stocks.

Again, a comparison reveals that a lot of money could have been made in just two days:

	Dow Jones	Percent Change
October 19, 1987	−508	−22.61
October 30, 1929	+186	+19.15

At least part of Black Monday's market bedlam was attributed to computer trading by institutional investors who program their computers to buy or sell stocks at a predetermined price.

After the 1987 crash, circuit breakers were installed to prevent large-scale, rapid, one-day market swings. Today we also have the Federal Reserve System poised to head off any beginning panic. A large service sector now exists that is less exposed to swings in the economic cycle than the manufacturing sector.

Alas, these two historical examples show that human nature never changes. Fear follows greed as surely as night follows day. This is what creates cycles. What they say is: "Those who don't learn from the past are bound to repeat it."

Still, when the Dow Jones Industrial Average is near its all-time high, luring speculators into the market, a "buy now before it goes up further" mentality prevails. When this occurs, other factors in the economy are affected as well.

WHAT CAUSES STOCK MARKET ANXIETY?

Stock market anxiety is often the result of other factors in the economy beyond the stock market. When inflation is very high, it destroys the integrity of the dollar and dollar-sensitive investments. Inflation causes people to lose confidence in the currency and they put their assets in real estate or gold.

When government spending is very high, it causes taxes to rise, making a new spiral upwards in inflation; prices go up, affecting our export business.

When inflation is low, growth in the U.S. economy will be low, affecting consumer spending and confidence. This will result in fewer job opportunities as companies cut back, lowering production and services.

High corporate and personal taxes cause the real return on investment and disposable income to go down.

When fear of recession (and possibly a depression) is growing, prices fall, purchasing is reduced, unemployment rises, inventories increase, and factories close. When business and personal debt is costly, more people will default on their loans in spite of moral obligation and legal liability.

When business and savings are low, consumer confidence falls. The money available for investment is scarce, which makes interest rates rise.

When interest rates are very high, the cost of using money will be high. Business expansion will be curtailed and consumer spending will go down.

When uncertainties are growing—threats of war, OPEC oil price increases, foreign competition with our industries, pollution and waste disposal, there will be growth in unemployment, social unrest, and crime.

During times of market anxiety there will be

• a probable drop in stock prices;

• low incentive to invest in common stocks; and

• low reward-to-risk ratios.

During times of market euphoria there will be

• a probable rise in stock prices;

- high incentive to invest in common stocks; and
- high reward-to-risk ratios.

There are two ways to learn anything—the hard way through harsh experience or the comfortable way by studying. Success in the stock market is dependent on the investor acquiring adequate knowledge and then using it with confidence.

The next chapter reviews the basics of investing.

Take fast hold of instruction; let her not go; keep her; for she is thy life.
Proverbs 4:13

Education does not mean teaching people to know what they do not know; it means teaching them to behave. John Ruskin

CHAPTER 3
INVEST WITH
CONFIDENCE

If you bet on a horse, that's gambling. If you bet you can make three spades, that's entertainment. If you bet cotton will go up three points, that's business. See the difference? Blackie Sherrode

Traders, banks, life insurance companies, mutual funds, and institutional investors are buying stocks for their portfolios. It's widely believed that the little guy doesn't have a chance. Yet, on the contrary, many smaller investors have found it possible to prosper handsomely in this challenging stock market. No matter whether stock prices are going up, down, or sideways, there are opportunities for profits.

The individual investor can also have great portfolio flexibility. Managers of institutional portfolios need to satisfy customer demands. They have to buy and sell large blocks of stocks to get in and out of the market; because the movement of large blocks of stock affects prices, they pay more to acquire stock and receive less when selling.

The average investor who deals in smaller amounts has more control of his portfolio. He can buy or sell stocks at any time, satisfying his own needs. In today's stock market, the individual investor has the flexibility to enter and exit the market at his own pace, without having to worry about market swings caused by portfolio managers.

UNDERSTANDING THE BASICS

Commit to Investing

Investing means committing capital with the expectation of growth or income. The challenge is to make this happen despite the hazards, some of which affect even "safe" investments such as bank accounts. Before investing, provide for emergencies with adequate savings and life insurance. The stock market is not for your first dollars. Provide for today first, then commit a portion of your current assets to investing.

Money falls into three categories: active, loafing, or dead.

Active money is invested to produce an after-tax rate of return greater than the current inflation rate. Active dollars are producing yet more dollars that also can be put to work. Active money makes use of the compounding principle; it is the foundation of a successful investment program.

Loafing money is working and earning some return, but after taxes the gains are not keeping up with inflation. Money in a conventional savings or checking account is a prime example of loafing money.

Dead money is exactly the opposite of active money. Dead dollars just sit around, not invested in anything and not producing any return.

Examine every dollar under your control and decide whether it is active, loafing, or dead. Most individuals have enough total dollars available to invest. A basic principle of successful investing is to convert the loafing and dead money into active money.

Dollar-Cost Averaging

Commit to investing a fixed dollar amount of money on a regular schedule. This is called *dollar-cost averaging*.

Dollar-cost averaging is the most reliable way to invest in the stock market because the average price per share is lower than the mean average price during the holding period. This is basic math: $100 buys 10 shares of a stock at $10, but only 5 shares at $20. The mean average price was $15. The investor who made these buys owns 15 shares and paid a total of $200, for an average price per share of $13.33.

Consider the following examples of dollar-cost averaging under different market scenarios:

1. In a Deciining Market:

	Investment	Share Price	Shares Acquired
	$300	$25.00	12
	$300	$15.00	20
	$300	$20.00	15
	$300	$10.00	30
	$300	$5.00	60
Totals	$1,500	$75.00	137

Average price per share ($75/5) = $15.00
Dollar-cost average per share ($1,500/137) = $10.95

This shows the importance of continuing your investment program throughout a declining market. The greatest number of shares were purchased when the share value dropped from $25 to $5. Any recovery above the dollar-cost average of $10.95 would establish a profit.

2. In a Flat Market:

	Investment	Share Price	Shares Acquired
	$300	$12.00	25
	$300	$15.00	20
	$300	$12.00	25
	$300	$15.00	20
	$300	$12.00	25
Totals	$1,500	$66.00	115

Average price per share ($66/5) = $13.20
Dollar-cost average per share ($1,500/115) = $13.04

Even in a flat market dollar-cost averaging can work to your advantage. As the example shows, the actual per share cost is 0.16 cents less that the average price of $13.20 per share.

3. In a Rising Market:

	Investment	Share Price	Shares Acquired
	$300	$5.00	60
	$300	$15.00	20
	$300	$10.00	30
	$300	$15.00	20
	$300	$25.00	12
Totals	$1,500	$70.00	142

Average price per share ($70.00/5) = $14.00
Dollar-cost average per share ($1,500/142) = $10.57

As the above example shows, the dollar-cost average per share of the five investments is $10.57 compared to the current $25 per share value.

The practice of dollar-cost averaging does not remove the possibility of loss when the market is below the average cost, but it clearly proves that the successful pursuit of the system will both lessen the amount of loss in a declining market and increase the opportunity for greater profit in a rising market.

Dollar-cost-averaging is an excellent strategy for the smaller investor just getting started; it also works well with IRAs and diversified mutual funds.

Dollar-cost-averaging also works well with Dividend Reinvestment Plans (DRPs—pronounced "drips"). DRPs are company plans that allow stockholders to automatically (usually monthly or quarterly) reinvest dividend payments in additional shares of the company's stock. Instead of receiving the normal dividend check, participating stockholders receive quarterly notification of shares purchased and shares held. Thus in DRP programs, the dividends reinvested increase the number of shares held.

DRPs are normally an inexpensive way of purchasing additional shares of stock. Some companies have DRP programs that permit participants to buy shares—including fractional shares if the cash is not available to purchase full shares—at discounts from market prices.

The investor relations office of a company will supply information about its DRP program.

Compounding—The Eighth Wonder of the World

Time is a factor that needs to be considered for investment success. Through time we realize the compounding effect on money.

Bernard Baruch, described compound interest as "the eighth wonder of the world." Compounding means that you not only earn interest on the principal, you then also earn interest on the interest. To see the effect of compounding over time on $100 invested using different rates of return, consider Table 3.1.

As the rate of return increases, the growth of an investment becomes impressive. The point is not rate of return but compounding. Despite the rate, compounding greatly increases the final return.

Risk *vs.* Reward: Finding Your Balance

Seasoned investors know that capital appreciation is linked to risk. Successful investors learn to take as few risks as possible and to make sure the risks they take are in keeping with the probable return. Usually the greater the investment risk, the greater the potential reward. Most of us have a risk tolerance influenced by personality and experienced.

Managing risk is a matter of understanding the calculation of risk and reward. Learn to minimize unnecessary risk and to balance risk and return. The successful investor takes only necessary risks.

Table 3.1: Compounding and Rate of Return on a $100 Investment

Rate of Return	Years Held	Value at End of Holding Period
6%	5	$134
	10	$181
	15	$245
	20	$331
10%	5	$164
	10	$270
	15	$445
	20	$732
15%	5	$210
	10	$444
	15	$935
	20	$1,971

Think about risk over the long term for any financial investment. Don't agonize by looking at the price quotes every day or even weekly. Make a long-term commitment and live with it.

There are two factors to consider when assessing your monetary risk tolerance:

1. *Capacity to invest.* Consider your job, age, marital status, children, and short- and long-term objectives.

2. *Risk personality.* Consider your willingness to accept uncertainty and perhaps loss of principal.

For practical purposes, everyone automatically feels that risk means a negative outcome—a loss. Yet investment risk is simply the probability that your return will be different, though not specifically higher or lower, from what you expect. Increasing risk means a chance of not only losing money but also making more than you expected. In our discussions we will use the term "risk" to mean simply the chance that an investment won't produce the expected return.

Several factors influence investment risk:

- *Inflation risk* has different effects on different types of investments. In times of increasing inflation the value of hard assets such as real estate, gold, and physical properties will increase dramatically. Inflation has the opposite effect on cash. During inflationary periods, cash loses value.

  When formulating any investment program, it is vital to project what the rate of inflation will be in the next five years. Doing this allows you to adjust your investment strategy periodically to minimize inflation risk.

- *Interest-rate risk* reflects the fact that short-term rates tend to respond to the supply and demand of money, crowd psychology, and government manipulation. Long-term interest rates tend to be governed by the prevailing inflation rate. A change of interest rates can have a profound effect on the value of investments designed primarily to yield income.

- *Business risk* is measured by the net earnings of a company. When earnings begin to head downward, the dividend and also the stock's price may decrease. When a company's earnings begin to rise, its stock will do well.

- *Market risk* is determined by competitive bidding—supply and demand.

- *Economic risk* can adversely affect valuations whenever there is a sudden change in the economy. For example, the increased number of bankruptcies in the early '90s were caused primarily by a general economic recession rather than a specific failing of the companies that went under.

- *Political risk.* Political forces are frequently irrational and can cause economic and financial chaos that affects companies or industries.

There are several ways to increase your risk tolerance:

Diversify. Distribute your assets among different types of investments. Because different financial markets gain and lose popularity at different times, you can take advantage of these swings by diversifying.

Place assets in *liquid* investments. Liquidity means that you can get in and out quickly if assets are soaring or declining. Liquid investments are stocks, bonds, and mutual funds. Non-liquid investments are real estate, annuities, and limited partnerships.

Take advantage of *market volatility*. Market volatility refers to rapid and extreme fluctuations of stock prices. This increases your risk exposure, but if you buy into the market a little at a time—have a systematic plan—this helps lower this risk. A long-term view is at least a five-year time span. Never panic when your net investment value goes down; most losses can be recovered with patience. Use market downs for adding to positions.

The Risk–Reward Relationship

Risk is actually lowest when people see it as the greatest, and when most people think of it as absent, it is actually the highest. Dick A. Stoken

In school the three Rs were reading, 'riting, and 'rithmetic. In the financial world the three Rs are risk, reward, and their relationship. Risk is directly related to reward. In other words, a larger reward can only be obtained by increasing your exposure to risk. If you undertake a more risky investment, you will be entitled to a higher rate of reward (return).

When assessing risk on any investment, have some standard by which to compare the rewards against the risks. Start with the most risk-free investment: the 91-day U.S. Treasury bill. The U.S. Treasury has never defaulted and unless the country completely collapses, will never default on paying the interest and principal on its obligations.
There are three risk principles:

1. Any investment should give a higher return than the current U.S. Treasury bill.
2. If two investments have the same rate of return, select the safer one for less risk.
3. If two investments have equal risk, select the investment with the higher return.

Investments in the stock market can be relative safe if the portfolio is diversified.

Diversifying

One way to increase risk tolerance is to diversify. Assets should be distributed among different types of investments because market sectors drift in and out of fashion.

There are about 30 core industries that take turns being in favor. We do know that a portfolio of diversified stock market industries will provide good returns with less uncertainty than a portfolio composed of a single investment or industry.

Diversification means different things to different people. One of the most popular approaches is called "asset allocation." Financial planners look at asset diversification in broad terms that include cash, fixed income, stocks, real estate, gold, etc. For example, insurance salespeople push insurance and annuities as good investments. Real estate salespeople sell income property. The largest asset of most people is the equity in their home. This value cannot be included in asset allocation.

To get started investing in the stock market, first list your investments in two categories:

1. Investments you control directly

2. Retirement assets (IRAs, Keoghs, 401Ks, etc.)

Many people forget that their retirement assets are an integral part of their investment program. Investment plans should be refined according to objectives. For instance, if your investment objective is income, you might own utility stocks. If it's total return, you would be less interested in high dividend from your stocks. Buying growth stocks says that dividends would not be of importance but appreciation in the value of the share price of the stock would be.

Once you have determined your current asset allocation and you're happy with the mix, you need to monitor your portfolio regularly. However, before you decide whether or not you're truly happy with your investment mix, ask yourself these questions.

- Do you currently have investments you do not understand and never did? This is the most common indicator of *adviser risk*—the result of acting on the advice of another individual. Rely more on yourself.

- Do you have money available to invest but keep putting off doing something about it? *Procrastination risk* results from lack of self-confidence and, perhaps, fear.

- Does your portfolio contain government-guaranteed bond funds? You are suffering from "it doesn't exist" risk. There is no such thing as a government-guaranteed bond fund. There are mutual funds that invest in government-guaranteed bonds, but the funds are not guaranteed.

- Does your portfolio contain stocks in only one sector or only the stock of your employer? Investing in individual stocks is supposed to be the American dream. However, by investing in only one stock or sector, you are taking on *non-diversification risk,* the risk of having too many eggs in one basket.

Most investors like to own both stocks and bonds. Stocks provide the action and promise of greater wealth in the future. Bonds provide a feeling of stability and an appealing flow of income. Does it really make sense to hold bonds in an investment portfolio?

We know that bonds show inferior returns to stocks and we also know that they aren't a very good hedge against declines in the stock market. So who needs bonds if stocks are usually moving up? No one wants an asset that provides lower returns than an alternative.

The thought that bonds are less risky is little justification for including them in a portfolio. There are better ways to hedge against the risk of owning equities than by holding bonds. In Section II of this book you will learn that you can hedge your risks in the market by using covered call options.

There is also a little noticed influence at work: Bonds are contracts that pay the same income, year in year out, until they mature. Stocks are claims on uncertain cash flows that investors buy because they expect higher dividend income in the future.

The long-term record tells us that bonds have a lower total return (income plus capital growth) than stocks. Since 1925, stocks have yielded a compound annual return of more than 10 percent, bonds a bit under 5 percent—a 2 to 1 advantage for stocks. I see no reason to hold bonds either as a hedge or for capital gains.

It is important to stay fully invested and to balance to your satisfaction the three investment objectives: growth, total return, and income. You can achieve all or any one of these objectives by investing only in stocks, in the shares of good companies whose earnings and revenues are growing at an accelerating rate. These companies should have a track record over many years that you can verify for yourself. This philosophy has helped true professional investors establish outstanding performance over the years.

A down market presents an excellent opportunity for investors to buy more shares of stock for their money, giving them a better position when the market comes back. An up market gives investors an excellent opportunity to sell shares of overvalued stocks for more money, giving them a better position when the market goes back down. Be in the right position for market swings.

How do we do this? Two stock market terms used to evaluate stocks are *Price/Earnings* (P/E) *Ratio* and *Dividend Yield.*

EVALUATING STOCKS

Price/Earnings (P/E) Ratio

If asked to select just one tool for evaluating a stock, most analysts would choose the price to earnings ratio. The P/E ratio simply describes the relationship between the stock price and the earnings per share. It is calculated by dividing the price of the stock by the earnings per share figure. For example, if the price of the share is $10 and the year's earnings of the company per share is $1, the P/E ratio is 10 ($10 divided by $1 per share). If earnings improve and the annual earnings per share is $2 and the price of the stock is still at $10, the P/E ratio will be 5 ($10 divided by $2 per share).

Individual P/E ratios and average P/E ratios can be found in the financial pages daily. These numbers give investors an idea of how much they are paying for a company's earnings. The higher the P/E, the more investors are paying and the more earnings growth they are expecting.

The P/E ratio is a critical piece of information because it shows the value of a stock in terms of how the company is performing rather than the selling price. Remember, however, there is no perfect P/E ratio. Some stocks that have lower earnings will have higher P/Es. These are usually growth stocks. An income stock, one that pays higher dividends, will tend to have a lower P/E.

The P/E, also known as the multiple, gives investors a single number that allows them to compare the values of stock prices. The higher a stock's P/E, the more investors are paying to share in the company's profits.

P/E ratios are driven by earnings expectations. Investors who are highly confident that a company will improve its profitability or remain profitable are willing to pay more for its shares, raising the P/E. If profits are weak or threatened, the P/E will most likely drop as investors demand compensation for the risk of disappointing returns.

Different P/E ratios are used to measure a company's prospects. There is the trailing P/E, which is the stock's current price divided by the company's reported earnings per share for the most recent four quarters. This is the most widely quoted figure in newspaper stock listings. It reflects actual results.

The estimated or forward P/E is derived by dividing a stock's current price by analysts' forecasts of future earnings per share for the next four quarters. These two P/Es may vary widely. The forward P/E is a projection, used by many brokers pushing a stock.

Dividend Yield

The dividend yield, often called simply yield, represents the annual percentage return the dividend provides. The yield of a stock is calculated by dividing the annual cash dividend per share by the price of the stock. For example, if a company pays its stockholders an annual cash dividend of $0.50 per share and the stock price is $10, the return, or dividend yield, is 5 percent ($0.50 per share divided by $10).

GENERAL INVESTMENT GUIDELINES

Can smart investors get rich by catching the wild swings between the highs and lows? Can the smart investor buy cheap and sell dear? A few really have, many lie about it, but more have lost money trying to time the market.

Smart individual investors can follow some simple guidelines, easily understood and researched, to obtain consistent returns.

Low P/E Investing

The low P/E investment strategy has worked well over time, providing investors with superior returns, though it rarely gives spectacular ones. Low P/E investors buy stocks with multiples below the stock market's average. For instance, if the S&P 500 is trading at 20 times last year's earnings and your stock sells for 10 times earnings or less, you have a low P/E stock.

Research studies from the 1930s up to now have validated the strategy of buying low P/E stocks. These low-priced stocks yield both higher dividend income and better capital gains. Who said you can't have both?

Low P/E value investing is generally the best strategy for all seasons; long-term results are impressive. A commentator on television recently said, "But aren't such companies out of favor with good reason? You have to wait forever for a turnaround." A good question, and one that has been asked of me many times. My reply is that you rarely have to wait forever. True, one should have the patience to wait, but markets make the wins come a lot faster than you might think.

If performance record is no secret, why doesn't everybody do it? Because most money managers and analysts believe earnings fit stock prices over time. Therefore, they favor stocks that look as if they will have rapid earnings growth for years to come and think the price of the stock will skyrocket.

That kind of future-gazing is dangerous because past performance has nothing to do with future performance. Past earnings growth is almost useless in predicting future earnings growth. The future is unpredictable, yet most analysts estimate from the past into the future and fearlessly forecast what is to be. Since they lean on such an undependable method, we should expect a high rate of error, especially when the forecasts are for extraordinary gains. Guess what? A high rate of error is what you find.

One delight of low P/E stocks is that they are so far down that just coming up to average creates substantial profits. If they go on from merely average to selling at a premium, the profits can be sensational.

An occasional freebie that comes with the low P/E strategy is the takeover play. Corporate treasurers on the hunt for value often find it precisely in this type of stock.

The low P/E strategy is not the only market strategy that works. However, it does produce above-average results on a consistent basis and at times can yield a good deal of excitement as well. Low P/E is not synonymous with dull.

Low P/E investors must learn to exercise discipline, discount rumors, hot tips, and fad stocks, and discount high P/E stocks.

Contrarian Investing

Go against the crowd, be a contrarian—buy when others are selling and sell when others are buying. When selling pressures are extreme, a contrarian buys and when buying pressures are extreme, a contrarian sells. The business-like concept of contrarianism is to watch the enthusiasm of the market and select investments that are sound but not currently in vogue.

The Crash of 1987 was a clear example of the trading rule: Buy when others are stampeding to sell. The selling stampede was over in less than a few hours, a combination of late Friday afternoon and Monday morning.

How can you take advantage of contrarian investing? No one can do your thinking for you. If you'd like to take advantage of contrarian investing, you must make the time to do the techniques outlined for you below and apply them to your contrarian stock choices.

Growth Stock Investing

Everyone dreams of finding a stock that appreciates by five to 10 times the original investment. Few have succeeded because they insist on investing for the home run. They seek the inside tip, the hot rumor, or they just trust stock selection to plain luck. Investing can be profitable but, like every other method of making money, it takes dedication, discipline, and hard work.

Basic to selecting stocks for growth is the discipline of maintaining a strict listing of financial and operating requirements. Once a stock is analyzed using the requirement screens, it may qualify for serious attention and possibly purchase. The following eight criteria should be used to screen potential growth stock candidates. Note, though, that it would be rare for a stock to show favorably in all eight categories. At this point your good judgment becomes the determining factor.

1. *Rising unit sales volume.* Leader or laggard? Is the company in a young industry or in an aged one? Look for new products and services being marketed. Over the past three years, the stock's price should have increased by a greater amount (total percentage) than inflation as measured by the Consumer Price Index (CPI) in the same period.

2. *Rising pre-tax profit margin.* There should be a rising trend in pre-tax margins over the past five years.

3. *Return on stockholder equity.* Does the company have the potential for an annual return of 15 percent on average stockholders' equity for the next two years? If the answer to this question is no, then invest your money elsewhere.

4. *Relative earnings per share growth.* Over the past three years the company's earnings per share should have advanced by a larger percentage increase than the calculated per share earnings of the Dow Jones Industrials.

5. *Dividends.* The company must pay a dividend and the dividend over the past five years should have risen steadily.

6. *Debt structure.* The company's long-term debt should be less than 35 percent of stockholders' equity. Debt of over 50 percent of stockholders' equity is a danger sign. The stock should be avoided.

7. *Institutional holdings.* Institutions should not own more than 10 percent of outstanding shares if the company is less than five years old. Ideally, in older companies institutional ownership will range between 10 percent and 30 percent. More than 40 percent of institutional ownership means the stock has been over exploited.

8. *Price/Earnings Ratio*. Appreciation in a growth stock results not only from steadily increasing per share earnings but also from a rising P/E ratio. The P/E ratio should relate to the rate of growth. A good rule of thumb: The P/E ratio should normally not exceed the projected growth rate of earnings.

Growth stock candidates should have favorable showings in at least five of these categories.

Dividend Income Investing

Selecting stocks for dependable current and rising dividend income requires different techniques than does growth investing. While earnings per share continue to be important, there must be more emphasis on the dividend record and financial stability.

A minimum of four selection criteria should be used to evaluate the stock:

1. A record of uninterrupted dividend payments for at least 10 years.

2. Increased dividends in at least five of the last 10 years.

3. Above average quality, demonstrated as a rating of A– or better by Standard & Poor's.

4. A sufficient number of shares available to provide liquidity in the marketplace: at least 10 million common shares outstanding with capitalization of at least $500 million. The larger the capitalization, the more secure the stock.

Putting It All Together

An efficient and personally rewarding investment plan uses a combination of all the investing strategies: Use fundamental analysis to select stock. Be contrary. Observe momentum. Buy and hold the stock until fundamental analysis and contrarian opinion say to sell the stock. Use all strategies to outperform the buy-and-hold method alone.

Where does all this leave you?

Follow the simple rules outlined below to determine when to buy, sell, or hold your stocks based on the above strategies.

When to Buy

Buy a stock when you have found one that meets the following criteria (Appendix I, Sources of Information, will help you locate the information you need to do your research).

1. Low P/E, no higher than half of the market multiple.
2. Current price less than the average of the 52-week high and low prices.
3. Increasing sales and earnings over the last five years.
4. Dividend pay-out ratio, the percentage of net earnings allocated to shareholders, that does not exceed 30 percent.
5. Informative and reassuring annual reports (see Appendix I to learn how to read these).
   Analyze the independent auditor's report.
   Verify all financial information and study all the footnotes.
   Read the CEO's report, which should state his opinion of the company's future outlook.

When to Sell

The criteria for selling, based on your research into the company's quarterly and annual reports, are the flip side of those for buying.

1. High P/E, at market multiple or beyond.
2. Current price higher than the average of the 52-week high and low prices.
3. Decreasing sales and earnings trends.
4. Net dividends that exceed net earnings.
5. Annual report no longer reassuring, provides negative information.

When to Hold

Hold a stock when it does not yet meet the criteria for selling, but certainly if two consecutive quarterly reports produce increasingly negative determinations, *sell*.

When negative news about a company is released, many investors immediately dump their stock. Be patient—hold. In most cases, the price will rise and the stock should be reevaluated; if you wouldn't buy the stock then, sell. Base decisions to sell on the guidelines you established for managing your portfolio. Do not sell based on emotions, because when the price of stock falls, you have already lost money.

Avoid this big mistake: Never wait to get even. Others are waiting, too. Whenever the stock price moves up, the others will start selling, keeping the price of the stock from rising further. When you decide to sell, don't make excuses to avoid selling. Make a decision and do it! Sell all or none of an issue. Don't hold back. When selling stocks, as in buying, have discipline and patience.

Buying and selling stock are easy. The hard part is knowing which stock to buy and when to sell it. Remember that every transaction must have two parties: For every buyer there must be a seller. Every time a trade takes place, two individuals have exactly opposite opinions of the future price of a stock.

BUILDING AN INVESTMENT PORTFOLIO

To obtain better performance out of your portfolio, consider the following guidelines:

- *Diversify by industry*. Assets should not be concentrated in too few or too many industries. A maximum of 10 percent for each industry is about right. Note the emphasis on industries rather than stocks.

  The entire list of industries and sub-industries is published periodically in the *Wall Street Journal* and *Barron's*.

- *Limit total holdings of a single stock*. In addition to industry diversification, each stock should initially represent no more than 10 percent of a portfolio. If a stock exceeds 20 percent, sell down to 10 percent.

- *Buy quality stocks*. Buying low-quality speculative stocks is only for the foolish investor. Beware of new, unseasoned stock issues. One rule of thumb is to always use the Standard & Poor's stock ratings. Never, except under exceptional circumstances, buy stocks rated lower than "A" (above average) by S&P.

- *Limit trading*. Frequent trading often means settling for small short-term gains. Further, it eliminates the opportunity for double or triple long-term gains. Small gains through trading also mean heavy commission costs and taxes. This eats into a portfolio's net profit.

- *Sell losers.* Consider selling whenever something significantly negative has occurred to the company that affects your original purchase decision.

STOCKBROKERS AND ADVISERS

Don't let them pick you (pun intended). Harvey C. Friedentag

Advisers

Investors who need advice should not seek it from a brokerage salesperson who works on commission. For a fee, you can get an unbiased opinion from a Registered Investment Adviser (RIA).

RIAs are monitored by the provisions of the Investment Advisers Act. The act, which is designed to protect investors from fraud or misrepresentation, requires disclosure of all conflicts of interest. Many RIAs specialize in a particular kind of investment; some will manage whole investment portfolios.

The Commissioned Broker

Stockbrokers these days are particularly anxious to build up their client rosters. Present or potential investors across the country constantly receive calls inviting them to receive special investment advice.

The more money a person has, the more appealing he is to brokers. However, even if you're not a big hitter, many brokers are still interested in handling your investments. The problem is finding the right stockbroker—and, unfortunately, we do need brokers in order to invest in the stock market.

Why do you need a stockbroker? Because only a broker can execute an order to buy or sell stocks. Only a broker whose brokerage firm is a member of a stock exchange can trade stocks on that exchange.

Stockbrokers are called registered representatives. This means that they have passed a Series 7 exam and are registered with the Securities and Exchange Commission (SEC) to represent their brokerage firms.

A specialist has the responsibility to see that there is an orderly market with no gross imbalances of buyers and sellers. For example, if for a brief period there is a shortage of buys or sells, the specialist either buys or sells shares from his personal account to keep the trading going.

If there is a large difference in the bid and ask price of the stock, the specialist may temporarily suspend trading until the imbalance corrects, either by the appearance of buyers and sellers or a change in the price of the stock. The system of brokers operating on the stock exchanges serves to maintain a very liquid and orderly market. That is why it is possible to buy or sell stock at any time.

As an individual investor you need the services of a broker to execute your trades, give you stock quotes, hold your certificates, credit and debit your transactions and dividends, and send you accurate monthly statements and annual reports. However, there are two types of stockbrokers: the commissioned broker and the discount broker.

Commissioned brokers are salespeople. They rarely have the time, experience, or documentation to form conclusions about an individual stock and they also tend to push the products sponsored by their brokerage firms.

Wall Street brokerage houses make the bulk of their profits from commission charges; from buying and selling. Many investments that look profitable show a loss when taxes and commissions are included. Factor these expenses into your calculations to determine whether you have a gain or a loss.

Brokers will always quote you their commission rate. However, all commissions are negotiable—ask for a better rate whenever you can.

Commissions are only part of the costs of investing. *Spreads* are hidden and can be as large or larger than the commission charge. A spread is the same as a mark-up in the retail business: It's the difference between an investor's selling price and a buyer's bid price. The spread can be a major influence on the results for both the buyer and seller.

How do spreads work? When you ask for a quotation on a security, you will be asked whether you are buying or selling. There are always two quotes: The bid is what you can sell the stock for; the ask is what you would have to pay to buy it. The difference between the two is the spread. The spread divided by the ask price is the percentage the spread contributes to your transaction costs.

For example, the quote for ABC Corporation is 4 1/2 bid and 5 asked; the spread is 1/2. This contributes 10 percent to the cost of buying ABC stock. This means that ABC has to go up 10 percent before you cover the cost of the spread and can start to earn the cost of the commission. If you use a discount broker charging a 3 percent commission, your stock has to go up 13 percent before you start making any money.

There are spreads on every security. The stock quote listed in the paper is the last trade price, generally in the middle of the spread. Spreads are necessary to have a market. Watch out for the large spread—the high percentage spread—which wipes out most of your potential profit before you even begin. Over-the-counter stocks (stocks not traded on a major exchange) have the largest spreads. Low-priced stocks have higher spreads than high-priced stocks.

Penny stocks (the very lowest-priced stocks) are the worst of all; the spread on some of these can be over 100 percent. You don't have to pay a commission on this type of stock because the commission is included in the quoted price.

Discount Brokers

It's true you need stockbrokers to buy and sell stocks, but you don't need their advice. Consider the following: The *Wall Street Journal* periodically publishes the investment performance of large investors, mutual funds, insurance companies, and banks. More than 75 percent of professional managers don't beat the market averages.

All the services provided by the commissioned broker are available from discount brokers. An investor can get commission discounts of up to 76 percent. What discount brokers won't do is give you advice, hold your hand, or try to sell you into or out of investments.

Trades with discount brokers are easy and fast and substantial commission discounts can make the difference between profit and loss on your securities transactions. Discount brokers can charge less for several reasons:

1. They do not have expensive research departments. You make your own decisions, so you don't pay for advice you don't need.

2. All account executives are salaried employees. This saves commission dollars.

3. Most large discount brokers are highly automated and use the latest in telecommunication technology. Their efficiency generates additional savings that get passed along to customers.

Discount brokers also offer the following:

- 24-hour order entry and low-cost margin loans.
- Assets usually protected up to $2,5000,000 through the Securities Investor Protection Corporation (SIPC).
- The ability to hold securities, which eliminates the risk and inconvenience of sending securities by mail.

As with commissioned brokers, discount brokers will notify you of rights offerings, advise on matters requiring stockholder action, send you proxies and financial reports, and, when there is activity in the account, send a comprehensive monthly statement.

Whichever type of broker you use, the instructions you provide will be entered on the computer and this will decide how your account gets handled (i.e., what will be done with your stock certificates, proceeds from sell transactions, and dividend and interest income). To get the best execution possible at the lowest commission, you need to know a few basics:

1. Help the person answering your request for quotes by supplying the symbol assigned to the stock. This eliminates errors and wasted time for both of you and improves the service. The representative answering the phone will ask you for your area code and telephone number. This information will be written on the order ticket and enables a prompt call back to report the execution price of your transaction. Today many brokerages have a sophisticated telecommunications system installed so you can call back a computerized system to get these messages; some systems will call you.

2. Initiate orders by saying you wish to *buy or sell*. Be prepared to provide your account number, security name, symbol, and the quantity you wish to buy or sell.

3. To assure an order gets executed at the opening price, enter an order at least half an hour before the market opens. The trading day opens at 9:30 a.m. and closes at 4 p.m. New York time.

4. When buying or selling, go in at the last price. This is the price of the last trade. Buy and sell at the last price. When you get quotes, ask for the last price. The last price is usually in the middle of the spread unless the stock is moving up or down, when the last price can be at the bid or ask.

   Trading at the last price minimizes the spread. Remember, if you buy with a 1/2 point spread and sell with a 1/2 point spread that is $1 in cost to you on a round trip. Trade with a discount broker, minimize the spread, and make as few trades as possible.

5. Also keep in mind the bottom line:
   Federal income tax   20% on long-term gain
   State income tax   10% on _____ gain

The spread	10% on the transaction
Commissions	3% on the transaction

Expenses are incurred whether you make a profit or not, though taxes only apply if there's a profit. Thus, a 10 percent loss on the market value of your securities could result in a 23 percent loss when the trade costs are figured in. If you sell the stock, you can figure in another 13 percent, for a total loss of 36 percent.

If you can cut your transaction costs by only 1 percent for each trade you make, you will save 2 percent from each round trip investment. If your performance would have averaged 10 percent a year, it will average 12 percent a year.

Applying time value tables is very helpful in this regard. If you invest $10,000 over 30 years at 10 percent, it will grow to $175,000. At 12 percent, your $10,000 will grow to $300,000. Don't let anyone tell you that commissions and spreads don't matter.

Transaction costs compared:

- Penny stocks have the highest fees: 20% to 300% (in and out)
- Limited partnerships fees are next: 6% to 20% (in and out)
- Secondary offerings fees are 5% to 10% (in and out)
- Full commission brokers charge 6%, unless you're a good customer who can haggle them down to 3% to 4% (in and out)
- Discount brokers get between 2% and 4% (in and out)
- Load mutual funds range from 4% to 8.5%. Most load mutual funds charge only on one end of the transaction, and competition from the no-load funds is forcing fees lower. Studies have shown that there is no correlation between a mutual fund's load and its performance.
- No-load mutual funds have no transaction costs. Watch for the difference between no-load and low-load. A true no-load has no sales or redemption charges. If you send in $1,000, you purchase $1,000 worth of shares at the current NAV (net asset value). All your money goes right to work for you. If you sell, you get all NAV back; there is no spread. The NAV is the same whether you are buying or selling.

Low-load funds have transaction costs. One little caveat: Many former no-load funds are becoming low-load funds.

All mutual funds have advisory fees or management fees to cover their services and expenses. These ongoing fees range from 0.5 percent to 2.5 percent of the assets under management per year. You can multiply the assets by the charges and see that this is no charity business.

All costs have to be considered when comparing the different approaches to investing. For example, if we have 100 shares of ABC selling for $20 and wish to switch to XYZ at $20, there will be two transactions with two commissions. In addition, there is a cost due to the spread. The bid-ask spread will increase the cost of switching. We may find in our example that the current spread on a $20 stock is 19 7/8 to 20 1/8: We would receive 19 7/8 for stock sold and pay 20 1/8 for the stock purchase, a difference of 25 cents.

6. Limit orders, which are orders to buy or sell only at a specified price, may be entered for the day only or can be placed GTC (Good-Until-Canceled). If you enter your order as a day order, it will automatically get canceled after the trading day if it did not fill. GTC orders remain in effect until you cancel the order or the order is filled. If the price of the security you wish to buy or sell moves dramatically away from the "Limit Price" on your GTC order, your order remains in effect and continues to be good until you cancel. You can cancel your limit order at any time. You are responsible for GTC orders; keep a record of them.

7. If you do not wish to accept an execution for part of your order at your limit price, you must specify "all or none" at the time you place your order. Limit orders placed all or none or with a minimum number of shares acceptable to the investor lose priority to market. Securities that you wish to purchase or sell can trade through your limit and you are not entitled to an execution because of the all or none or the minimum amount restrictions placed on your order. By instructing all or none, you are avoiding the possibility of multiple fills and commissions.

8. For reporting purposes, the security industry consolidated all transactions. Various financial periodicals incorporate the trading ranges for all trading that occurs in all markets for each individual security on any given day. Be aware that some limit orders placed may not get executed even though the price range of the composite market printed in your newspaper is beyond the limit. This can

occur because your security has not traded through your limit price on the exchanges where your order was placed. The brokerage reserves the right to choose the exchange when listed securities trade on more than one exchange.

9. All NYSE-listed odd-lot orders (orders for less than 100 shares, a round lot) received for common stocks one half hour or more before the market opens will be executed at the opening price without an odd-lot charge added or subtracted.

   All other odd-lot orders for common stock on the NYSE will be charged an eighth of a point odd-lot execution charge by the odd-lot specialists that execute the order. Preferred stock odd-lot charges are variable.

   On the AMEX, all odd-lot orders in stocks that trade below $40 a share are charged an eighth-of-a-point execution charge. All odd-lot orders in stocks that trade above $40 a share are charged a quarter-of-a-point execution charge.

10. Orders placed on the NYSE are left to the discretion of the specialist who executes the order. Round-lot orders with odd lots attached that get entered as a market order will generally get executed at the same round-lot price.

    Round-lot orders with odd lots attached should be entered odd-lot-on-sale. Otherwise the round lot will be executed but the odd lot will not. This is because the stock did not trade an eighth of a point through your limit to effect the execution of the odd lot. When the odd lot does not execute on the same day as the round-lot shares you must pay a full commission when your remaining odd-lot shares get executed. Since for commission purposes brokerages combine the round and odd lots together, odd-lot orders entered odd-lot-on-sale will be executed on the same day as the round lot and the commission charge will be based on the total shares executed that day.

11. Avoid delays in receiving payment and avoid unnecessary handling of certificates by having the brokerage hold your securities in street name and hold your funds for safekeeping. Street name describes securities held in the name of a broker for a customer. As long as the brokerage is a member of the SIPC, your account is protected. Ask for the official brochure. The brochure explains the purposes of the corporation and the amounts of protection offered. Funds left on deposit will earn market rate interest.

Many investors leave the proceeds from sales transactions in their account, where they earn interest until the funds are reinvested. You can call your brokerage at any time to ask that a check be sent for all or part of your funds left on deposit.

If you prefer to receive payment for securities sold, you can expect a check to be issued on settlement day or the day following receipt of securities sold, whichever is later. Settlement day is three trading days after the transaction.

If you prefer to receive securities (certificates) registered in your name, expect to receive them about four weeks after settlement date. Securities purchased are not transferred until your payment is received.

12. Endorsing stock certificates:
    * Do not sign in spaces indicated as blank.
    * Give your brokerage firm a power of attorney in the space on the back of the certificate. This renders your endorsed certificate non-negotiable until the firm releases the power with a stamp and a signature.
    * Date the certificates.
    * On the reverse side of the certificate, sign your name exactly as it appears on the face. If your certificate(s) are in the name of two or more parties, all must sign.
    * When you mail securities or checks to your brokerage firm, include a copy of your trade confirmation or write your account number on the check or stock certificate.

13. You must pay for securities bought and deliver securities sold by the settlement date of your transactions. If your funds or securities are not received in time, the brokerage may be forced, pursuant to Regulation T of the Board of Governors of the Federal Reserve System, to sell out securities purchased or buy back securities sold to cover any unsettled portion. You are liable for any deficit incurred by the broker on your behalf as well as for the commission for both transactions. If for any reason you are unable to meet the settlement date, call your broker and ask for an extension.

14. When you have an account problem, call the brokerage firm after the market closes and, after you have made your request, ask for a return call to be informed about the status of your problem and the proposed action for its resolution. Remember the name of the individual handling your request.

15. As a customer of a discount broker, from time to time you will receive a special reduced rate offer to buy a variety of financial magazines, computer software, books, and newspapers. Take advantage of the ones you are interested in.

RULES TO INVEST BY

Common sense is the knack of seeing things as they are, and doing things as they ought to be done. C.E. Stowe

Some investors will pay $20 for an investment worth $10 because it seems sure to go to $30 in the near future. An investment may or may not improve. In order to safeguard against guesses, remember the following:

- Be patient—remember the value of time.
- Always try to be fully invested in common stocks.
- Let your stock advance at least 50 percent before selling.
- Hold a stock four to five years.
- Don't invest in the following:
  Gold, silver, or other precious metals
  Real estate
  Collectibles
  Penny stocks
  New issues
  Foreign stocks
- Don't tie your investment to the industrial bond rate.
- Don't tie your investment decision to computer analysis.
- Consider past earnings growth as an important factor.
- Estimates of future growth in earnings are important. Do the research necessary to estimate a company's future earnings growth.
- Invest in turnaround companies only if in-depth research shows that future earnings prospects justify it.
- When you decide to get out of a stock because of bad news, sell it at-the-market to make sure you sell. At-the-market means that the broker has authority to sell 100 shares at the best price available. Use market orders only when you absolutely want a transaction completed.

- When you decide to buy a stock, set a price that the stock is worth to you. Use a limit order to make sure you do not pay more. When placing a limit order, you will be asked, "How long is the order good?" The order may be in effect one day or may be good until canceled. Limit orders are like a market order except that the broker has to make the trade at a previously specified price or better.

- Place a stop-loss order not because you want to sell but because you want to protect yourself if the stock drops rapidly in price. Stop-loss orders are limit orders to sell; they are insurance. Many successful investors use stop-loss orders routinely, placing them 5 percent to 20 percent below the current stock price. As the stock rises in value, new stops are entered 5 percent to 20 percent below the new current stock price. A stop-loss order is an investment strategy since it eliminates the possibility of taking big losses.

  Stop-loss orders do have some serious problems: Normally stock prices fluctuate and you don't want to be in the position of selling your stock simply because it temporarily dipped in price. The reason for the dip is of utmost importance in determining what action to take: buy, hold, or sell. Another problem is the possibility of large down days, as in the 1987 crash. Stocks fell so rapidly that the price was below the stop-loss limit before a sale could be made. A stop-loss order in this situation is like a limit order. When they got to your order the price was already lower, so you did not get an execution.

  Therefore, when using stop-loss orders, keep them to yourself as mental orders, not on file with your broker. This will keep your commissions down.

- Use the vast quantity and quality of information at your disposal to guide your decision process. You will have a head full of useful facts and a head start on most of your competition.

- All common stocks you buy or hold should stress safety. Once you have built up a portfolio, you can take some prudent risk.

- Buy stock in large companies. Ignore those clowns who tell you that the only stocks to own are small company stocks. Large companies suffer less accounting gimmickry than smaller companies and results are spotted more quickly.

- To properly diversify a portfolio, own stock in at least 10 different companies and own no more than two companies in one industry.

- Consider dividend yields a strong plus. A dividend yield is not all but it is an important contributor to total return. A yield also supports the stock in bad markets. The dividend is an excellent indicator of the stock. An increase or decrease tells you how the company is doing.

- Always reinvest your dividends and capital gains in order to add to your portfolio's worth.

- Scout for low P/E companies. Try to distinguish those suffering temporary setbacks from those that are in a long-term downtrend. In low P/E investing, you have a lot going for you, but it won't produce fast results.

- Sell a stock when the P/E greatly exceeds the market average and replace it with another low P/E stock.

- Look for financial soundness and a low debt-to-equity ratio. A low debt-to-equity ratio suggests conservative financing and low risk. Accept a high debt-to-equity ratio only if inflation rates are high.

- Look for high book value. Book value can be a guide in selecting under-priced stocks.

- Look for upward price movement and good earnings trends. Consider past stability in earnings growth as of some importance.

- Look for good companies whose stock has been so badly beaten down that it has nowhere to go but up.

- Use a discount broker. If you need advice, use a Registered Investment Adviser.

- Never buy stocks at market; use limit orders.

- Ask and obtain answers to the following questions:

  1. What is the state of business and the economy? Where are we in the business cycle? Is the boom likely to top out? Are we in a recession?

  2. What is the state of the market? Are we in the early stages of a bull market? Has the low point of a bear market been reached? Is the bull market about to top out?

  3. If answers to the preceding questions seem favorable, what industries are likely to grow most rapidly? Are there any special factors that favor a particular industry?

  4. Which company or companies within the industry are likely to do best? Which companies are to be avoided because of poor prospects?

- Finally, dispel the following investment myths:

Myth #1: The stock market is for gamblers.
Gambling? The stock market is not gambling. You have control. At a horse race you can buy the favorite; when the bell rings all you can do is watch the race, win or lose. You're locked in. In the stock market you can, whenever you wish, get off your pick if it's losing and get on a winner. With games of chance you most certainly can't do that.
While there is an element of risk in investing in the stock market, it's far from gambling. Do your homework, follow sound investment principles, and have patience. Long term, the stock market can be the best available way to increase your wealth.

Myth #2: Buy low and sell high to get rich.
The myth is true—but unfortunately, achieving it is practically impossible, because what is the low? and what is the high? Most people who try to get rich this way usually end up making their broker rich by generating commissions, with little profit for themselves.

Myth #3: The stock market is for those who have money to lose.
A variation of the first fable, this one implies you should have money to lose to be in the stock market. This is not true. Everyone can benefit from common stock investing, including individuals who can only afford to invest a small amount each month.

Myth #4: Individual investors cannot afford to compete with the professionals.
This is utter nonsense. Individual investors not only can compete with professionals, they often outperform them. Earl Gottschalk once noted in the *Wall Street Journal* "Your Money Matters" column (January 2, 1990), "You have an advantage over professionals who only have a given amount of time in a day to keep track of many stocks for several clients. You have the opportunity to complete a more in-depth study of your stocks. Taking note of companies close to home and tracking their management is a good start."

Myth #5: The stock has gone down so much, it has to be a good buy now.
If the stock price has gone down dramatically, there has to be a reason. If the reason is poor fundamentals or discouraging prospects for the future, the stock may be far from a good buy at

any price. Don't send good money after bad just because the price is lower. Study the fundamentals and if things have changed for the worse, admit the mistake. Then consider what your money could be doing in another investment.

Keep in mind that a stock that drops in price is not automatically a loser. It could drop for reasons that have little to do with the company's underlying value. When a stock moves contrary to the market or its industry, try to find out why.

Myth #6: I'll wait to get even before I get out.

This could be a big mistake. You may never get out. Don't hesitate to remove losers from your portfolio. It doesn't mean you are doing a poor job of investing; everyone has them. If you're right 80 percent of the time, you will be a successful investor. Admit your errors and go on.

Myth #7: It takes a lot of money to invest.

It doesn't take a lot of money; it takes a lot of time. Whatever sum of money you choose to invest in the stock market regularly, through good times and bad, will lead to unbelievable results over long periods of time. The biggest mistake for many people is not realizing the magic of compounding and the reinvestment of dividends.

Purchase of an odd lot, less than 100 shares, costs a little more. For most issues, you will pay an additional eighth of a point (12 1/2 cents) a share to odd-lot brokers for trading a small amount of stock. Beyond that, brokers generally do not charge higher commissions for such trades. A small odd lot of stock may run up against the minimum commission at many brokerage houses. If you're buying shares that you plan to hold for five years, the total effect of a slightly higher price will be small.

Some brokerage houses offer a Monthly Investment Plan where you can start investing. Once you purchase the initial shares, you can add to them at regular intervals. Your money plus dividends will buy more shares.

Myth #8: Don't buy during a bear market.

This is another big mistake, one made by amateurs and professionals alike. A bear market can be the best time to buy good quality stocks at bargain prices. If after studying a stock, you find it has good fundamentals, a bear market will insure that it represents a good purchase value as well. Many investors rely on a contrarian strategy. During out-of-favor periods, they buy undervalued shares that appear low priced compared to the worth of the company's assets. Then they wait for other investors to recognize the stocks' true worth and bid up their prices. That is the basis of sound value investing.

Myth #9: You can't go broke taking a profit.

No, you can't go broke, but you sure can miss the top dollars. The greatest investment success stories come from investors who have bought and held, keeping stock in companies that have continued to grow year after year, market cycle after market cycle, sometimes for decades. If you only hold your stocks for the short term you might limit your profits. While short-term profits range from 5 to 25 percent, holding for the long term could earn two, three or four times the price paid for the stock.

In addition, those who sell have to pay taxes on the profits.

Myth #10: What my stocks are doing daily is important.

This is why the current Dow Jones Industrial Average has such prominence in daily newspapers and on the nightly news report. Yes, it's important to keep up-to-date on the stocks in your portfolio, but when the price dips next Tuesday, you needn't panic. When you're holding for the long term, the daily rises and falls of the stock market should be of no more than passing interest to you. More important are the fundamentals underlying the stocks in your portfolio. Are they still strong? What are the company's plans? Is there anything on the positive or negative side that may change your outlook? Read the annual report and other sources of information to keep current on what the company is doing. There is little in the daily news that is significant to the wealth you can accumulate over the long term.

Myth #11: If interest rates go down, stocks have to go up.

When the federal discount rate gets cut, stocks have to go up because savers are pulling their money out of low-interest savings accounts and looking for something better. I hear people say, "I'm not happy with the interest rate I'm getting at the bank, so I'm going to buy stocks." This will not help if the fundamentals of the stocks in your portfolio are not strong.

Myth #12: Stock is scarce.

All a company needs to create more stock is a printing press.

Myth #13: Soaring profits are around the corner.

What if future earnings don't happen? In a really bullish market that's no problem. Last year an analyst said a stock was low priced at $15, a mere 40 times the 88 cents a share it was going to earn. The company's earnings turned out to be minus $2.61. "Don't give it a thought," the analyst said, "The loss was from a write-off, and write-offs don't matter. The stock is now at $11 and is cheap against this year's earnings." There's always a gap between forecasts and reality. Don't bet on what analysts think might happen in the future.

Myth #14: The stock market is rational.

If this were true, investors could buy blindly, knowing that any company's price on Wall Street does reflect its prospects and risks.

What can a rational investor do? Buy stocks low when interest rates are high and all the experts say, "Stocks just can't compete with bonds." But did you have the guts to buy in those days? If you didn't, you shouldn't be buying stocks today.

The old-fashioned reality in the stock market is based on long-term fundamentals. Undoubtedly, stocks do move too fast, both up and down, because of the instant widespread distribution of facts, rumors, and false information. Yet, when stocks are cheap—not just looking cheap but truly being cheap, based on earnings, cash flow, equity values, and dividends—they should be bought.

The remainder of this book explains, in greater depth, these techniques, practices, and strategies using covered call options. Hold on to your money until you have a working knowledge of what follows. It is essential that you have a full understanding of the principles discussed in order to take advantage of my investment strategy.

Facts do not cease to exist just because they are ignored. Aldous Huxley

Past experience should be a guidepost, not a hitching post. Anonymous

SECTION II
OPTIONS

CHAPTER 4
OPTIONS AND
THE STOCK MARKET

Money is a sixth sense which makes it possible for us to enjoy the other five. Richard Ney

You are now ready to be introduced to the real world of the stock market. A stock can do three things: it can go up in value, it can go down, or it can stay the same.

Regardless of what the stock does, it can be to your advantage. You will learn how to deal with this volatility and make money. You will learn why options can be the versatile tool for risk management and yield enhancement. And you will learn about the brokerage industry's wonder weapon: Exchange-listed options.

The real beauty of options is their versatility. Many investors consider them a highly speculative investment. In fact, the greatest benefit of options is in reducing and defining risk, not increasing it.

WHAT IS AN OPTION?

"Option: 1. to wish, desire. 2. . . . choosing; choice. 3. the power, right, or liberty of choosing. 4. the right, acquired for a consideration, to buy or sell . . . something at a fixed price . . . within a specified time." *Webster's New World Dictionary.*

An option in the stock market is a legal contract that gives the right to buy or sell a specified stock at a specified price—the *strike price*—before a specified date—the *expiration date*.

An option to purchase stock is a *call* and an option to sell stock is a *put*. A call or a put represents 100 shares of stock.

Listed options are securities that are regulated by the exchange on which they are traded. An option contract is for 100 shares (unless adjusted for stock splits or stock dividends).

An option buyer (the holder) pays a premium for the right to buy or sell the underlying security. The seller (the writer) of a call option is obligated to sell the underlying security to the option buyer if the call is exercised. The writer of a put option is obligated to buy the underlying stock if the put is exercised.

The essential elements of an option contract are the *strike price*, the *premium*, and the *expiration date*.

The strike price, or exercise price, is the price at which the underlying security can be bought or sold.

The premium is the price the buyer pays in return for the rights conveyed in the option.

The expiration date is the last day on which the option can be exercised.

OPTIONS AND COMMON STOCKS

Options share many similarities with common stocks. Both options and stocks are listed securities. Orders to buy and sell options are handled through brokers in the same way as orders to buy and sell stocks. Listed option orders are executed on the trading floors of national Securities and Exchange Commission (SEC)-regulated exchanges where all trading is conducted in an open, competitive auction market.

Like stocks, options trade with buyers making bids and sellers asking. With stocks, the bids and asks are for shares of stock. In options, the bids and asks are for the right to buy or sell 100 shares (per option contract) of the underlying stock at a given price per share for a given period of time.

HISTORY OF OPTIONS

The Bible recorded the first option short sale when Esau, for a mess of pottage, sold short to his brother Jacob the inheritance he had not yet received.

In the year 1694 put and call options were introduced in London, England. Three centuries later put and call options continue to be important security dealings.

Introduced into this country about a century ago, put and call options soon became a favorite speculative tool of the old-time Wolves of Wall Street. The conventional shied away from them, and many to this day cast a prejudiced eye on options. The ordinary investor used to regard dealing in puts and calls as a special, complicated maneuver, tinged with evil.

About 40 years ago, when the SEC threatened to stop options trading, the Put and Call Brokers and Dealers Association instituted rules and standards that resulted in a degree of respectability for put and call options. Since then, stock options have become better understood and more widely used by investors to hedge against price movements, to protect unrealized profits, and for potential tax savings.

Although the history of trading options extends over several centuries, it was not until 1973 that standardized, exchange-listed, and government-regulated options became available. In only a few years, these options almost displaced the limited trading in over-the-counter options. Option trading has become an indispensable tool for the securities industry.

FUNCTIONS OF THE OPTIONS CLEARING CORPORATION (OCC)

Standardized option contracts provide orderly, efficient, and liquid option markets.

Options are an extremely versatile investment tool. Because of their unique risk/reward structure, options can be used with other financial instruments to create a hedged position.

A stock option allows you to fix the price, for a specific period of time, at which you can sell 100 shares of stock. For a price (premium) this option is granted. Unlike other investments where the risk may have no limit, options offer a known risk to buyers.

The Options Clearing Corporation (OCC) selects companies to be listed on the option exchanges. Though most companies favor this listing because it adds interest in their securities, the decision to list is not theirs. Listing requirements state that trading of the company stock must be at a high volume and above $10 per share. If these requirements are not met, new option expirations are not traded; when existing contracts expire, companies not meeting the requirements are delisted.

The OCC guarantees that the terms of an option contract will be honored. There are no ifs, ands, or buts with options.

Before the existence of option exchanges and the OCC, an option-holder who wanted to exercise an option depended on the ethical and financial integrity of the writer or his brokerage firm for performance. Also, there was no convenient means of closing out one's position before the expiration of the contract. The OCC, as the common clearing entity for all SEC-regulated option transactions, resolves these difficulties. Once OCC is satisfied that there are matching orders from a buyer and a seller, it severs the link between the parties. In effect, OCC becomes the buyer to the seller and the seller to the buyer, guaranteeing contract performance. The seller can thus buy back the same option he has written, closing out the transaction and terminating his obligation to deliver the underlying stock. This in no way affects the right of the original buyer to sell, hold, or exercise his option. All premium and settlement payments are made to and paid by the OCC.

OPTION VALUATION

The premium of an option depends on:

1. The price movement (volatility) of the underlying stock;
2. The time to expiration of the option (more time = more money); and
3. The difference between the current stock price and the strike price.

Volatility is a measure of stock price fluctuations. The ideal stock for option writing would be one with medium volatility and growth potential based on solid fundamental value. A dividend is preferable but not a requirement.

The underlying security is the stock which can be purchased or sold according to the terms of the option contract and is the base on which a sound option writing program rests.

An Example: ABC Corp. stock is selling today on the NYSE at $32 a share. The option buyer has several choices:

- Choice one gives the buyer the right to buy ABC Corp. stock at $25 a share.

- Choice two gives the buyer the right to buy ABC Corp. stock at $35 a share.

Choice one is more valuable, since the option buyer would rather have an option to pay $25 for a $32 stock. As a result, it costs more to buy choice one than to buy choice two.

Sellers know this, so as the stock price rises and falls, the option price rises and falls with it. As time elapses toward the expiration date, the option price falls because the time value is eroding (wasting). Option investors, like stock investors, can follow price movements, trading volume, and other pertinent information day by day or even minute by minute.

There are some important differences between options and common stocks that should be noted, however. Unlike common stock, an option has a limited life. Common stocks can be held indefinitely in the hope that their value will increase, while an option has an expiration date. If an option is not closed out (*exercised*) before its expiration date, it ceases to exist as a financial instrument; thus an option is considered a wasting asset.

There is not a fixed number of options as there is with common stock shares. An option is simply a contract between a buyer willing to pay a price to obtain certain rights and a seller willing to grant these rights in return for the price. Thus, unlike shares of common stock, the number of outstanding option contracts, commonly called *open interest*, depends solely on the number of buyers and sellers interested in receiving and giving these rights.

Unlike stocks that have certificates, option positions are shown on printed statements listing buyers or sellers by a brokerage firm. This procedure sharply reduces paperwork and delays.

Finally, while stock ownership gives the holder a share of the company, including certain voting rights and rights to any dividends, option holders benefit only from price movement in the stock.

THE OPTION CONTRACT

The *option contract* has the following elements: type (a call or a put), underlying security (deliverable security), strike price, and expiration date. All option contracts that are the same type, covering the same underlying security and having the same strike price and expiration date are referred to as an *option series*; they are *fungible*, meaning interchangeable.

Fungibility is a very important term for options traders. All options in an option series are fungible, as are such assets as commodities and securities. For example, an investor's shares of ABC Corp. left in custody at a brokerage firm are freely mixed with other shares of ABC Corp. Likewise, stock options are freely interchangeable among investors, just as wheat stored in a grain elevator is not specifically identified as to its ownership. Later in the book you will learn how to use fungibility to your advantage.

THE OPTION PREMIUM

The *option premium* is the cash price exchanged when options are bought and sold. Premiums fluctuate depending on the duration of the contract, the strike price, and the current price of the underlying stock. Premiums can run as high as 25 percent of the value of the underlying stock. For example, for a volatile stock selling at 20 ($2,000 for 100 shares), the premium for a call to be exercised in nine months might be 5 ($500) when the exercise price is also 20. Shorter-term options on more stable stocks carry smaller premiums, from 2 percent for expiration in a month to 10 percent for those with longer maturities.

The writer of an option is obligated to deliver the underlying security if the option is exercised. Whether or not an option is exercised, the writer keeps the premium.

Premiums are quoted on a per share basis. Thus, a premium of 1 represents a premium payment of $100 per option contract ($1.00 x 100 shares). Premiums are quoted in points and fractions.

An option buyer cannot lose more than the price of the option premium. The option will expire as worthless if the conditions for profitable exercise do not occur by the expiration date. Remember, options are a wasting assets that go to zero with time.

In an option contract the premium is the only variable. The number of shares, the expiration month, and the strike prices are all standardized.

THE COVERED CALL WRITING STRATEGY

To some investors, the very mention of the word "option" evokes images of very speculative, highly leveraged trades. However, there is an option strategy that is quite conservative and appropriate for most equity investors.

Many of us have seen advertisements that state: "Buying an option on stock offers the chance for unlimited profit with a limited risk." Don't believe it. Buying options is like buying lottery tickets. True, you could win big, but the odds are against your winning at all.

If buying options is a risky endeavor, how about selling them?

Selling options is not usually risky and you win small, but often. The people making money in options trading are those who sell them; their winnings are never spectacular, but they are regular.

The option selling strategy is called *covered call writing*. The covered call writer either:

1. Buys stock and simultaneously sells an equivalent number of covered call options (buy-write); or
2. Sells covered calls on stock that is already owned.

   The benefits of this strategy are three-fold:
   * It places money in the investor's account.
   * It increases the investor's probability of profit, often substantially.
   * It allows investors to make a profit, sometimes 50 to 80 percent annualized, and not worry about swings in stock prices.

A *covered call option* is sold by an investor who owns the underlying stock. In case the covered call option is exercised, the seller is covered by the stock owned. An investor who holds 100 shares of AT&T common is considered covered if he writes one option on his stock.

From now on we will refer to covered call options simply as options. These options are the most popular and widely used.

In our strategy we will always be selling, which means we will always be taking in money. The buyer is on the risky side: He has to predetermine the expected value of the stock at the time period he wants. The buyer is always paying, which means he is giving money away. We will always be taking money in.

Since we will always be selling covered call options, we become the *covered call option writer*. Option writers are normally conservative investors seeking additional current income.

We will buy an optionable stock and write covered call options on that stock. An optionable stock is one listed on an option exchange. Currently there are about 1,500 optionable stocks.

We will use only exchange-listed options because they enhance returns in a variety of ways:

* by hedging downside risk;
* by combining risk protection with upside potential;

- by allowing for tactical adjustments without the need to buy or sell securities; and

- by increasing liquidity through more risk/reward alternatives.

Of my own investing, 90 percent is in the stocks of high quality companies; 10 percent is in slightly higher-risk companies with the potential of higher rewards. Of my gains, 75 percent result from dealing in covered call options.

I'm conservative. As the saying goes, I'm concerned less with the return on my money than I am with the return of my money. I wish to stress that nobody knows for sure if a stock is going to go up or down. I do know the prices at which I am a happy buyer or a happy seller. If the stocks do nothing, I'll be getting more than the return on a money market investment.

Covered call options allow me to own stock in a more conservative manner. Instead of trying for the big capital gain, I can get a good return, with some participation on the upside. I know what the rates of return will be if the stock is "called" away or if it stays unchanged through expiration. If the stock goes down, the covered call option premium offers some cushion.

The key is discipline.

Discipline fortifies the heart with virtuous principles, enlightens the mind with useful knowledge, and furnishes it with enjoyment from within itself. Hugh Blair

GETTING STARTED CORRECTLY

Understanding the fundamentals of covered call option writing is essential in learning how to distinguish a good covered call write. A call option gives its holder the right to buy an underlying security. Call options are thus derivative securities because their values are derived in part from the value and characteristics of the underlying security. For example, the XYZ Corp. May 20 call option entitles the holder to buy 100 shares of XYZ Corp. common stock at $20 per share at any time before the expiration date in May.

When you write (sell) covered call options, you start with an immediate, sure, limited profit rather than an uncertain potential greater gain. The most you can make is the premium you receive, even if the price of the stock soars. When you write covered call options on the stock you own however, any loss of the value of the stock will be reduced by the amount of the premium received.

To become proficient covered call sellers and to reduce the likelihood of choosing the wrong price, become familiar with some of the terminology relating to covered call selling. Though the nomenclature in the options industry can be intimidating, there are useful terms active call writers should be aware of. These include

- strike price
- at-the-money
- in-the-money
- out-of-the-money

Strike Price

The *strike price* is the price per share that the holder of the option must pay to buy the corresponding stock if he chooses to exercise that right.

The strike price for a call option is initially set at a price close to the current share price of the underlying security. Subsequent strike prices are set at the following intervals: 2½ points when the strike price to be set is $25 or less; 5 points when the strike price is over $25. New strike prices begin when the price of the underlying security rises to the highest or falls to the lowest strike price currently available.

If the strike price of a call option is less than the current market price of the underlying security, the option is said to be *in-the-money*. If the strike price of an option is more than the stock price, it is *out-of-the-money*. If the strike price equals the current market price, the option is *at-the-money*.

At-the-Money

At-the-money covered call options are written at an exercise price that is at or close to the current price of the stock.

For example, in January, Paul Prudent buys 100 shares of Always Good (AG) at 20 and sells a July covered call option at the strike price of 20 for $2 ($200 on a 100-share contract). Paul realizes that AG may move above $22 in the next six months but is willing to accept the $2 per share cash.

John, the option buyer, who expects that AG will move well above $22, gets the right to buy the stock for $20 at any time before the expiration date in July.

Paul will not realize a dollar loss until the price of AG goes below $18. At $22 the profit starts for John (the buyer). What happens if AG stock rises to $30? At any time before expiration in July, John can exercise his option and pay $2,000 for stock then worth $3,000. After deducting about $300 (the $200 premium plus commissions), John will have a net profit of about $700. Paul will have a $2 gain on the stock plus the $2 option premium, for a total gain (less commissions) of $400—a 20 percent gain in six months on his $2,000 investment.

If the price of AG stock moves up to only $22, John can call (exercise the option) and buy the stock, but will not be even because of the premium and commission costs. Paul would have sold the AG stock for $22 ($20 for the stock and $2 for the option), plus two dividends at $25 each.

If the price of AG stock stays at $20, John will not exercise, Paul still owns the stock and keeps the $200 premium, and he can write a new covered call option.

In-the-Money

In-the-money is a more aggressive technique that requires more attention but can bring in greater profits and tax benefits.

In-the-money covered call options are sold at strike prices below the current price of the stock. Since the options involve a greater money premium, there is a higher percentage of return and, in a down market, more protection from loss.

Assume that in January, Joe Smart buys 300 shares of Fantastic Furs (FF) at $20 ($6,000) and sells three July 15 ($15 is the strike price) covered call options at $7 each ($2,100). If FF stock stays at or below 15, Joe keeps the premiums and the stock. If the stock goes to 25, he can buy back the options for $10 (stock price $25 – strike price $15 = $10), or a total price of $3,000 ($10 x 300) to set up a short-term tax loss of –$900 ($3,000 buyback – $2100 option money = –$900). Joe can sell the shares for $7,500 ($25 per share x 300 shares = $7500), for a $1,500 gain ($7,500 sale price – $6,000 cost)—a $900 option loss for a net gain of $600.

If FF stock drops below 15, Joe keeps the premiums and writes new covered call options. He won't suffer a loss until his real cost of $13 ($20 stock cost – $7 option premium) drops to $13.

In this example, Joe started with $6,000 worth of stock but had an actual cash outlay of only $3900, because of the option premium received.

Out-of-the-Money

Out-of-the-money options are written at a strike price that is above the current price of the stock.

Sally Surething owned 300 shares of Safest Safes Industries (SSI) which she had bought at $10 two years earlier. Sally, anticipating a rise in the stock price, wrote covered call options at a higher strike price: In January, when SSI was at $20, she wrote three option contracts for July 25 at $3 ($900 premium: $3 x 300 = $900). In July SSI was at 26 and the option was exercised. Sally received $25 for each share for a total sale price, counting the $3 option premium, of $28 while the stock was only at $26.

Out-of-the-money covered call options work best in an up market with quality stocks bought when undervalued and there are longer-term options (six to nine months) with a rich premium. Of course, if the stock price was below $25 at expiration, the option owner would not exercise his rights and Sally could still have the $900 and can rewrite the covered call options again for a new premium.

A Note of Caution

If the holder of a call option decides to exercise his right, he must notify his broker. The brokerage firm, upon receiving an exercise notice, will assign one or more of its option-writer customers, either randomly or on a first-in, first-out basis. That is, they may assign the total to one seller or they may assign part of several sellers' contracts to fill their obligation. Regardless of the assignment, call option writers are subject to the event each day that some or all of their short options may be assigned. This results in multiple commissions for the brokerage firm and you should object strongly if this occurs. The assigned writer remains obligated, however, to sell the underlying shares of stock at the specified strike price.

CASH PREMIUM = TIME VALUE + INTRINSIC VALUE

The *cash premium* equals the sum of the *time value* and the *intrinsic value*. The cash premium received from the sale of the covered call effectively reduces the cost of buying the stock (though not the cost basis). This provides a measure of downside protection for the investor against declines in the stock price. The premium effectively increases the yield on the investor's position.

Intrinsic value is the in-the-money portion of a call option's price—the difference between the exercise price or strike price of the call option and the market value of the underlying security.

Time value represents the value of the time remaining to expiration of the call option. Generally, investors want to sell calls with as much time value as possible. Time value is the part of a call option's total price that exceeds intrinsic value. If the call option has no intrinsic value, it consists entirely of time value. In general, the greater the time value on a covered call, the better the downside protection and the greater the total return.

The most important consideration when developing a covered call write strategy is the underlying stock. Investors must have a positive opinion of the stock. Look for stocks with solid value; forget about second-guessing the market to find a stock that is going to be a big gainer.

Investors must also consider the rate of return generated by the position. The combined return from stock appreciation and the option premium and dividends received should exceed alternative investments with comparable risk/reward characteristics.

This is a conservative investment approach. The idea behind it is to earn added income from assets that are solid and productive in their own right. The stock selection process used is a defensive one. Stocks chosen are those that pay a good call option premium. The emphasis in this approach is on not losing.

By following a rigid discipline, many errors can be avoided. There is absolutely no room for the latest hot tip or gut feeling. To repeat: The most important part of the investment program is the quality of the underlying security.

The stocks included in your covered call options program should be evaluated using the criteria discussed in Section I. Review dividend yield, the P/E ratio, and the stock's market history. The stocks in your portfolio will be chosen from those that meet your criteria because they are optionable at a high premium.

Also, remember to diversify as much as possible. Normally, the more stocks in a portfolio, the lower the risk.

For the covered call option strategy to work, take these things into account:

1. Buy a carefully selected stock.

2. Sell covered call options (collect a premium).

3. If the option expires, write another option (collect a new premium).

4. If the option is exercised, sell the stock for the strike price.

The goal of covered call option writing is to reduce risk and to generate extra income. *Changing Times* (now called *Kiplinger's Personal Finance*) said (March, 1986), "After wincing over single-digit returns, it's no wonder that interest-starved investors perk up at talk of squeezing extra yield from their stocks." This extra yield is after commissions and on top of dividends paid by the stock.

The fascinating question is why would someone be willing to give you extra profit? The answer is there are really two "someones" involved.

The most visible someone is the brokerage firm offering the chance for riches by enticing investors to buy call options. The firm is willing to offer the chance for unlimited wealth, because it gets a commission to act as broker.

The other someone is the seller of the covered call option. Most sellers are professional traders, both on and off the exchange-trading floor. They are willing to sell call options for the same reason as insurance companies and mutual funds—in order to increase their earnings and have a protective hedge against a downside in the market.

They are familiar with the markets; they know the market will move and they can balance potential risk against potential reward. There are always opportunities for individual investors to use covered call options as a part of a total plan. With careful selections and constant monitoring, selling covered call options can:

- Boost annual income by 15 percent or more.

- Provide tax benefits and cost less (commissions are small compared to those of stocks).

- Give a variety of choice (in underlying assets, strike prices, and time frames).

But remember, never buy options. Sell them! Buying options is speculation; selling options is investing!

AVOID THE OPTION PREMIUM TRAP

Many investors hunting the option premium become elated by the high potential returns available on writes of volatile stocks like take-over issues and "fad" stocks. The time premium for call options on these stocks is very high, making them tempting to write. However, economic reasoning offers little support for initiating writes on extremely volatile stocks.

Profit from any dramatic gain in the price of the underlying security may be limited to the strike price of the call option sold. However, participation in losses higher than net proceeds from selling the option is *not* limited.

In short, there is no option premium big enough to protect you from a downside break in a volatile situation. An investor who initiates a call option write on a $20 stock because of the "fat" $4 premium that could be earned by selling an option can see the stock price drop to $7 as the market sees through the phantom fundamentals propping up the price of the stock.

Stick to fundamental value and never write on a stock you wouldn't feel comfortable owning at the net price paid for the optionable stock.

The option page in the financial section of major newspapers reveals a world of stocks from which to choose. The best way to choose may be the process of elimination. Don't select stocks with fundamental characteristics you do not find attractive. From the rest of the optionable stocks, choose a diversified list of fundamentally sound companies that can be analyzed thoroughly and monitored easily. This will enhance familiarity with a company's earnings and a stock's trading patterns, therefore minimizing surprises.

THE OPTION SELLER'S ADVANTAGE

Every dollar that call option buyers lose goes into the pockets of call option sellers. Since a majority of options expire worthlessly, it means the seller of a call option who waits for it to expire has a good chance of making a profit. Investors with adequate time and the financial resources to operate on the seller's side of the options market can build a fortune with these odds.

Writing (selling) covered call options turns off most uneducated investors for a couple of reasons. To begin with, there doesn't seem to be much money in it. Most call options transactions sell for $500 or less. Assuming the call option expires totally worthless, the option writer can't earn more than the premium received for the option. Option writing doesn't deliver the instant returns of 200 percent or 300 percent that are possible when you buy options.

However, over the long term the option writer will make far bigger profits than will the speculator who buys call options in hopes of making a quick killing. In the options market, option writers have 75 to 85 percent odds in their favor. It's the time premium that tips the odds in favor of the call option writer.

While buying call options offers the promise and the occasional reality of huge profits, selling call options is the wonder weapon that holds the potential for getting interest-free loans, avoiding the concerns of price swings in the stock market, and getting a 25 to 40 percent annualized yield. Selling call options puts money into an investor's account.

Why is something this good kept almost a secret? Because it's more complicated than just buying and holding stocks, and most brokers don't really understand the full potential.

Many individual investors want to write call options and reap the benefits. Since the option writer is selling time value, he can frequently make money whether the market stays the same, goes up, or goes down. This increases the probability of profit.

USING MARGIN

Margin is the amount a customer deposits with a broker when borrowing from the broker to buy securities. A brokerage account that permits an investor to buy securities on credit and to borrow on securities already in the account is a *margin account*. Buying on credit and borrowing are subject to standards established by the Federal Reserve and by the firm carrying the account. Interest is charged on any borrowed funds for the period that the loan is outstanding.

In the following paragraphs, we will briefly discuss why investors should use *margin* to enhance their investment portfolio and ease taxes. Margin is discussed in more detail in Chapter 6.

At this point you might be wondering why I'm advocating what you see as a dangerous method. You ask:

Options? That's gambling.

Margin? That's borrowing.

Why are you suggesting that I borrow to gamble?

Even some of Wall Street's savvy investors recoil at the mention of options. Yet, as I've shown, covered call options at least can be a part of a very conservative hedging strategy.

Other savvy investors point out the poor performance of the mutual funds that have used options in their strategy: "If the pros can't do it, you can't do it." Yet others will tell you that the stock market is dangerous enough and it's no place to use borrowed funds to invest.

It's important to understand the true meaning of margin. The rational use of margin need be no more risky than using a mortgage to buy a house.

Ideas about options and margin are the most prevalent misconceptions in the stock market. In the following chapters you'll see what can be done with options and margin. I'll also advise you about mistakes to avoid and how to protect against the misuse of options.

Options and margin can be effective tools for enhancing the total return on an undervalued, diversified, long-term stock portfolio. Not many stock market participants (investors, stockbrokers, bankers, lawyers, and accountants) really understand all the technicalities of margined and optioned portfolios.

Using margin can increase the percentage of profit on an investment by a surprisingly large amount, so you can have a larger portfolio and take advantage of downturns in the market to buy bargains. The cost of borrowing on margin is very low. Within certain limits, margin loan interest can be tax-deductible as well.

A margin loan against the value of a portfolio is limited to 50 percent. Since it is open-ended, there are no monthly payments, though interest is deducted from the account monthly.

WRITING COVERED CALL OPTIONS = PRUDENT INVESTING

Covered call option selling has now become so established as a prudent, conservative method of investing that it has received official recognition by almost every regulatory body that has control over investments. The Comptroller of the Currency, who regulates national banks, has ruled that writing covered call options is an appropriate activity for bank trust departments.

Insurance Commissioners of most states have now ruled insurance companies may write covered call options on some of their own investments. Various officials entrusted with seeing that pension plans are properly administered have given their blessings to the writing of covered call options for pension plans. In addition, an enormous number of conservative, professionally managed investment groups, including churches, university and college endowment funds, and union welfare plans, have begun to write covered call options. Clearly, the time has come when the speculative stain of options should be completely and finally removed from the concept of writing covered call options.

By its very definition, the word "conservative" means being more concerned about protecting what one has than with increasing its value through changes in market prices. Writing covered call options protects the investor from declines in the stock price up to the amount of the premium from the sale of the call options. Therefore, using covered call options with stock ownership is more conservative than simple ownership of the stock.

Selling covered call options is one of the few forms of investing where you can compute exactly what the return on investment will be if the position is successful. True, when an investor buys a stock, he can compute the dividend, but the results of that investment one year after purchase are going to be determined mostly by what has happened to that stock's price.

With the covered call option strategy the idea is to continue to own common stock and also to get some downside protection, taking some profit if the stock stays still or moves up.

Everyone who owns securities should understand and use covered call options. They can provide extra income, set up tax losses, make possible protective hedges, and usually limit losses. However, to make money with call options, you must work hard, research thoroughly, review often, and follow strict rules.

Why bother with options at all? In case you haven't noticed, there is risk in stock ownership. Harvey C. Friedentag

Time is what we want most, but what alas! we use worst. William Penn

CHAPTER 5
COVERED CALL
OPTIONS AND
POTENTIAL RETURNS

Hindsight is always 20:20. Billy Wilder

Currently, there are approximately 1,500 optionable stocks and more are pending. With so many choices, how can you decide which ones are best for your portfolio?

Let's start with the well-known KISS strategy—Keep it Simple, Sweetheart—and go on from there.

Keep it simple. Don't sell uncovered call options because they are free money—the stock can't go up that far. This could produce losses.

Keep it safe. Don't exceed your risk tolerance.

Keep it sensible. Just because you read about some alleged mastermind making a killing with call options doesn't mean you should try to copy his operation. Don't try anything that will take you beyond your eating and sleeping comfort points.

Keep it diversified. Gain greater safety through buying different stocks rather than large trades of a single stock—diversify.

Finally, and most important:

Keep it disciplined. Losses sometimes come because of the wrong stock, the wrong strategy, the wrong timing, or just bad luck. Often losses get larger because of a lack of discipline. When you implement a covered call option strategy:

1. Set an approximate goal—the point where you expect the strategy to produce profits.

2. Establish an exit point in case the trade goes against you.

Above all, avoid naked writing. *Naked writing* is selling an option on something you don't own. It's very risky because you must always be ready to buy the stock and then immediately sell it to the option buyer on demand—no matter how high the stock has risen in price. Avoid this very high gamble.

Covered (not naked) call options can be used in their protective capacity as instruments for the transfer or reduction of risk. Keep in mind that writing covered calls against your stock holdings is safer than just holding the stock. The concept is to continue owning common stock yet provide downside protection and take some profit if the stock stays still or moves up. This approach is not as glamorous as the more risky strategy of buying calls. However, our goal remains to enhance and preserve capital.

STARTING A COVERED CALL PROGRAM

From now on we will refer to covered call options simply as *covered calls*.

If you already own 100 shares or more of an optionable stock, your decisions are already partly made for you. Since there are always three time frames available, your first question is probably whether you should sell the three-month, six-month, or nine-month covered call.

The way to decide is to compare the average monthly return for each time period. The shortest time frame pays the highest return on a monthly basis. The longer-term covered call offers a commission savings that should be considered. Also, you may prefer longer-term covered calls to avoid the necessity of multiple transactions for the same time period.

The Pricing of Covered Calls

The most important factors that contribute value to a covered call contract and influence the cash premium at which it's traded are

- the price of the underlying stock (compared to the strike price); and
- the time remaining until expiration (the time value).

If the underlying stock price is in-the-money, there is intrinsic value. For example, if a covered call's strike price is $25 and the stock is trading at $35, the covered call has an intrinsic value of $10.

If the underlying stock price is at-the-money or out-of-the-money, there is no intrinsic value.

For in-the-money covered calls the time value premium is the difference between the strike price and the intrinsic value. For at-the-money and out-of-the-money covered calls the time value premium is the covered call premium.

These values are in dollars per share. For example, a covered call contract covering 100 shares of common stock when trading on the options exchange at 2 points has a total value of $200 (100 shares x $2 = $200).

The intrinsic value is always a positive number or zero; it equals the stock price minus the exercise price. The time value of a covered call premium is the difference between the dollar value of the premium and the intrinsic value in dollars of the covered call contract. Thus, time value equals the premium value minus the intrinsic value.

The longer the time remaining until a covered call expires, the higher the premium. This is because more time means a greater possibility that the underlying share price might move to make the covered call in-the-money.

The time value of a covered call does not decrease at a linear rate. The time value falls off gradually until close to expiration, and then falls off rapidly. Consider the following examples.

Example 1: In-the-Money
On May 21 (expiration date):
ABC Corp. common stock closed at 35
ABC Corp. August 30 covered call closed at 8

Intrinsic value (35 – 30)   = 5
Time value   (8 – 5)   = 3
Total premium   8

Covered call contracts have a maximum life of nine months. During this period the premium can vary widely, from very low to very high. The value may be small when the time to expiration is very short or when few option buyers expect that the market price of the underlying common stock will rise before expiration. The value will be high when the time to expiration is long, or when the stock is above the strike price.

Example 2: At-the-Money
On May 21:
ABC Corp. common stock closed at 35
ABC Corp. August 35 covered call closed at 4½

Intrinsic value	= 0
Time value	= 4½
Total premium	4½

On August 20th, when ABC is valued at \$32.50, the option holder will not call your 100 shares of ABC. You keep the \$450 premium money (pre-tax) and write another option on your ABC Corp. stock.

When the market price of the common stock is below or less than the strike price of the covered call contract, the intrinsic value is zero. In this instance, any value of the premium is entirely time value.

Example 3: Out-of-the-Money
On May 21:
ABC Corp. common stock closed at 35
ABC Corp. August 40 option closed at 3

Intrinsic value	= 0
Time value	= 3
Total premium	3

On August 20th, with ABC valued at \$32.50, the option holder will not call your 100 shares of ABC. You keep the \$300 premium money and write another option on your ABC Corp. stock.

Results:
The Example 1 option holder calls 100 shares of ABC from you at \$30:

100 ABC Corp. sold at	30
Premium received	8
Total received (per share)	38

Example 1: In-the-Money	Sell the stock, make \$300.
Example 2: At-the-Money	Keep \$450 premium, write another option.
Example 3: Out-of-the-Money	Keep \$300 premium, write another option.

The stock price when we did the option was $35 a share. On the August 20th expiration date, the stock price is $32.50.

May 21:	100 ABC Corp. at $35.00	$3,500.00 stock value.
August 20:	100 ABC Corp. at $32.50	$3,250.00 stock value.

If we have to sell at $30	$3,000.00
Plus option money	800.00
Took in (on hand)	3,800.00
Minus stock value	3,500.00
Option net profit	$300.00

Your net profit is $300. This was the time value of the cash premium of $800 you received three months previously. Your net profit can never exceed the time value of the covered call you have sold.

Other factors that give covered calls value and therefore affect the premium are volatility, dividends, and interest rates.

Volatility refers to the frequent, large price fluctuations of a stock. This volatility of the underlying share price influences the covered call cash premium. The higher the volatility, the higher the cash premium.

Dividends: Cash dividends paid go to the stock owner. Cash dividends affect covered call premiums through their effect on the underlying share price. Covered calls reflect stock dividends and stock splits because the number of shares represented by each covered call is adjusted to take these changes into consideration (e.g., one option at $40 becomes two options at $20).

Higher *interest rates* have tended to result in higher covered call premiums. That portion of the time value attributable to the interest rate factor will be greater.

DETERMINING THE BEST TIME VALUE

In determining how long a covered call to write, remember that usually a three-month covered call option will have a higher average monthly return than will a six- or nine-month option. If your analysis gives no indication of a stock price rise or fall, follow the rule of selling the covered call that gives you the highest return on a monthly basis.

However, if your analysis indicates the stock price will be lower in three months, it would be to your advantage to write a nine-month contract. In this situation, as you'll see below, the nine-month contract will

produce less premium than writing three consecutive three-month contracts.

If your analysis indicates that the stock price will increase in three
months, it will be to your advantage to write only a three-month contract
and, on its expiration write another three-month contract, which would
total more premium than an original nine-month contract. For example:

Stock price on May 21 $50.00	3-mo.*	6-mo.**	9-mo.***
Aug 50 option premiums	3.00		
Nov 50 option premiums	3.00	5.00	
Feb 50 option premiums	3.00	2.50	6.50
9-month total premiums	9.00	7.50	6.50

Note: Only half of the second six-month premium is included in the total. We
sold two six-month options in the first 12 months, but are figuring it for just nine
months: one full six-month option and half of the second six-month options.

*Sell 3-month option three times; total premium: $900.

**Sell 6-month option one and a half times (one 6-month and half of second
6-month income for 9 months); total premium: $750.

***Sell one 9-month option for $6.50 income; total premium: $650.

If your analysis indicates that in three months the stock may be down 5 and
selling at 45, sell the long-term option.

Stock price on May 21 $50.00	3-mo.*	9-mo.
Aug 50 option premiums	3.00	6.50
Nov 50 option premiums	1.00	
Feb 50 option premiums*	1.00	____
9-month total premiums	5.00	6.50

*Stock price at $45.00.

When the stock price is 45, covered calls with a strike price of 45
probably would have a premium for three months of 2½. You could sell
this covered call after the first option expired, but the risk is that if the
price goes up to 50, to buy back the $45 option for $5 you would incur a
$2.50 loss. Selling the Nov 50 covered call and getting a 6 percent return
in three months, equaling 24 percent a year, is hardly losing. *Note:* We are
using averages and will expect to lose on 15 percent of our trades.

If you believe that your stock is likely to go up during the next three
months, write the shortest-term covered call. If the stock does as well as

you expected, you will be assigned, sell your stock, and be free to reinvest in another optionable security.

Remember, your analysis gives no indication of a stock price rise or fall, you can follow the rule of selling the covered call that gives you the highest return on a monthly basis.

DETERMINING THE BEST STRIKE PRICE

What are the results of choosing a high strike price rather than a low one? Since your stock has covered calls available at three or more strike prices, which one should you write?

Covered calls with the highest strike price will be the most profitable if the stock goes up; those with the lowest strike price will perform the best if the stock goes down.

A lower strike price offers a higher premium. A strike price below the stock price provides a premium including intrinsic value, which gives downside protection. For instance, when ABC Corp. was selling at 39, its five-month covered calls were priced as follows:

Strike price	Premium
35	$5 \, ^3/_4$
40	$2 \, ^5/_8$
45	1

When ABC stock remains at 39 until the covered call expires, the writer of the 35 call option will realize $35 from the exercise of the option (the sale of the stock) plus the $5.75 premium, making a total of $40.75. The seller of the 40 strike price will keep his stock and the premium of $2.62. The seller of the 45 strike price covered call will keep his stock and the premium of $1.

If the price of the stock had risen to 45, the seller of the 45 covered call would have done best. The seller of the 35 covered call would have $40.75 as above. The seller of the 40 covered call would receive $40 for his stock plus the $2.62 option premium, making a total of $42.62. The seller of the 45 covered call would have the value of his stock, $45.00, plus the premium of $1, a total of $46. It would appear that the best strategy is to sell the covered call with the highest strike price.

But what if the stock had gone down? Suppose the price of the stock fell from 39 to 34. None of the covered calls would have been exercised, and each seller would keep both the stock and his premium. Now, each covered call seller has stock worth $34, and the 35 strike price writer has a premium of $5.75 in his account, the writer of the 40 strike price covered call has $2.62, and the one who did best in the previous case ends up with just $1.

Which strike price, then, to write?

To answer this question, return to the basic reason for selling covered calls: The seller of covered calls is willing to give up a large potential future gain—which may never happen—in exchange for a sure profit now, which is the option premium.

The main reason we are selling covered calls is to obtain an assured income. The one big risk is that the price of the underlying stock will fall. Therefore, my general rule is

When selling a covered call, give first preference to the call option with the lowest strike price.

This rule provides the most protection against a decline in the stock's price. If the price of the stock rises, you would have been better off selling at a higher strike price, but you're still making a profit with the lower strike price. When you sell a covered call option, you're giving up possible occasional future home runs for a steady stream of base hits.

Common sense is the knack of seeing things as they are, and doing things as they ought to be done. Calvin Stowe

In the ABC Corp. example, when you sell the covered call with the $35 strike price you've made an extra $1.75, 4.4 percent, on your stock. Since this profit will be earned in five months, it means that you're earning an annual rate of 10 percent. This is great when you consider that the buyer of your call option is giving you free insurance for every dollar that your stock could fall to 35.

If you strongly believe that the stock has a good expectation of rising, you may wish to sell at the highest strike price. You get to keep part of the rise in your stock and only turn over to the option buyer the excess increase. If the price goes down, remember you will be left with a loss on your stock and only a small premium to cushion that.

Each investor in stock has an opinion about what return on investment is acceptable based on his own financial condition and the amount of risk he is will to assume. Covered call selling is no different. A covered call writer accepts a certain amount of risk due to the potential volatility of stock ownership. Therefore, the likely returns from a covered call sale must exceed those that might be earned on a totally risk-free investment (if there was such a thing).

One general guideline for setting minimum acceptable returns on a covered call sale is to seek potential returns of at least double the selected risk-free rate on an if-exercised basis.

Note: Attractive potential returns can be found in utilities and some major oil companies. The reason for the attractive return is the dividend. But, because these stocks have low volatility, cash premiums are small, so an option premium would add only marginal income. In such situations, it would be more sensible to buy the stock and own it without selling a covered call. The cash premium provides little in the way of income or downside protection, yet it will limit the upside potential.

SELECTING THE BEST UNDERLYING STOCK

If you do not own any optionable stock and would like to sell a covered call, you have to decide which stock to buy.

First, select a stock that you personally will want to own and hold, using all the information given to you in the first part of this book. For instance, if you decide that there will soon be a downturn in drug stock profits and prices, there is no point in buying Merck just because it may have a high option premium. Select a stock that you would want to own if you were not going to be selling covered calls. If the price of the stock declines and the call option is not exercised, you will still own the stock.

Second, you will have to check the quotations to find the option premiums for the various stocks. The amount of the option premium varies depending upon many factors. Select a stock whose covered call has a high cash premium. Usually the lower-priced stocks have the higher premiums (percentage-wise).

Third, pay attention to the strike prices for the stock you're considering. As suggested earlier in this chapter, the safest covered call is the one with the lowest strike price. You're seeking less risk; pick a stock whose covered call has a strike price below the current market price of the stock. If after doing all your research you're optimistic, you can pick a stock that has a strike price well above its current price, understanding that this will reduce the amount of your premium and provide little downside protection.

While checking current strike prices it's good practice to check not only the current call option time period, but also the next three-month period and the one after that, because not all strike prices may be available for future periods. If the price of the stock has declined substantially, the highest strike price is not going to be offered for the longer covered calls. If the price of the stock has increased substantially, the higher strike price may not yet have opened for trading.

DETERMINING WHEN TO TAKE ACTION

Once you've written a covered call, wait until it expires and then decide which call option to write next. In most cases, though, it will be to your advantage to decide whether or not there are changes that could be made before expiration to enhance your profit or decrease your downside risk, or perhaps both. Watch your positions to see if you're earning the maximum amount from the time value of the covered calls you have sold. (Adjusting your position is explained in more detail in Chapter 8.)

Two situations should alert you to action:

1. *The current price of any covered call you have sold has declined to a small fraction of its original premium.* If any of your outstanding covered calls are selling for 0.25 or less, it's time to act. No matter what happens, the maximum remaining profit you can make from that covered call is only 25 cents a share and your downside protection is also only 25 cents a share, which is nil. The low price results from time decay or a decline in the price of your underlying stock.

2. *The time value of your outstanding in-the-money covered calls has fallen to a small amount.* When this amount begins to approach zero, there is no reason to continue the position. You still have the possibility of loss if the stock declines, but you no longer have the opportunity of a meaningful profit.

UNDERSTANDING PUBLISHED OPTION TABLES

Premiums (prices) for exchange-traded options are published daily in many newspapers. Many investors have difficulty in understanding option tables. *Investor's Business Daily* has introduced an alphabetical option listing that includes the volume of contracts. The *Wall Street Journal* has also started publishing an improved option listing that appears to be the best now available.

A portion of an option table for a trading day is shown in Table 5-1. Though this is only a small part of the total information, it's sufficient for us to understand the listings. It also will allow us to introduce some common option words.

Let's look more closely at the listing for ABC Corp. options. Although we will not be using puts, it's very important to realize where they get listed so as to never confuse the cash premiums for different types of options.

Table 5-1: *Wall Street Journal* Format

Option/Strike	Vol	Exch	Last	Net Chg	a-Close	Open Int
ABC Jun 12½	60	CB	3½	+½	16	72
ABC Jun 15	200	CB	1	+¼	16	570
ABC Jun 17½ p	40	CB	1½	-¼	16	660
ABC Jul 15	800	CB	1½	+¼	16	600
ABC Jul 15 p	80	CB	¼	-_	16	120
ABC Jul 17½	200	CB	¼	...	16	1000
ABC Aug 15	1300	CB	1¾	+¾	16	408
ABC Aug 15 p	200	CB	½	-_	16	930
ABC Aug 17½	10	CB	½	...	16	812
ABC Aug 20 p	4	CB	4	-¼	16	25

p-Put   a-Stock close

The published table reflects the previous day's trading. Under Option/Strike is the name of the underlying security, the expiration month, and the available strike price.

Vol (volume) is the number of trades.
Exch is the exchange on which the option is traded.
Last is the closing price of the option contract.
Net Chg (Net Change) is the difference between the last trading price from one day to the next.
a-Close is the closing price of the underlying security.
Open Int (Open Interest) the number of options outstanding.

Table 5-2 shows a format used by a local paper that also reflects the previous day's trading.

Table 5-2: Local Paper Format

Options & NY Close	Strike Price	Jun	Calls-Last July	Aug	Jun	Puts-Last July	Aug
ABC Corp	12½	3½	r*	r	r	r	r
16	15	1_	1½	1¾	r	¼	½
16	17½	r	¼	½	1½	r	r
16	20	s**	s	r	s	s	4

* r = no option trades that day

** s = no such option exists

In the first column is the name of the underlying security and its clos-ing price.

The second column lists the available strike prices.

The next three columns (Calls-Last) show the closing premium for each of the closing months for which calls are trading.

The last three columns (Puts-Last) show the closing premium for the each of the closing months in which puts are trading.

In Table 5-1, the in-the-money ABC Corp. August 15 calls closed at 1¾, or $175 per contract.

The out-of-the money ABC Corp. August 17½ calls closed at ½ or $50 per contract.

Though for purposes of illustration, commission and transaction costs and tax considerations are omitted, these factors will definitely affect a strategy's potential outcome, profit or loss, on your income tax return.

UNDERSTANDING EXPIRATION CYCLES

A specific optionable stock trades in only one of three cycles. Each cycle is composed of four three-month periods, as follows:

Cycle 1	Jan	Apr	Jul	Oct
Cycle 2	Feb	May	Aug	Nov
Cycle 3	Mar	Jun	Sep	Dec

Maximum option life is about nine months. Expiration dates are the third Friday of the month. New nine-month periods commence on the Monday following the third Friday of the month. Options can be traded for the time remaining to their expiration. In addition to the cycles, prior to expiration, trades can be made for a period from one day to the end of the following month. For instance, on the third Friday in January until 4:15 p.m. (EST), a trade covering a period from one day through the end of February is possible. This creates a new short-term option expiration, as seen in Table 5-3.

Table 5-3: Expiration Cycles

January Cycle

Expiring Month	Available Months			
Jan	Feb	Mar	Apr	May
Feb	Mar	Apr	Jul	Oct
Mar	Apr	May	Jul	Aug
Apr	May	Jun	Jul	Oct
May	Jun	Jul	Oct	Jan
Jun	Jul	Aug	Oct	Jan
Jul	Aug	Sept	Oct	Jan
Aug	Sep	Oct	Jan	Apr
Sep	Oct	Nov	Jan	Apr
Oct	Nov	Dec	Jan	Apr
Nov	Dec	Jan	Apr	Jul
Dec	Jan	Feb	Apr	Jul

February Cycle

Expiring Month	Available Months			
Jan	Feb	Mar	May	Aug
Feb	Mar	Apr	May	Aug
Mar	Apr	May	Aug	Nov
Apr	May	Jun	Aug	Nov
May	Jun	Jul	Aug	Nov
Jun	Jul	Aug	Nov	Feb
Jul	Aug	Sep	Nov	Feb
Aug	Sep	Oct	Nov	Feb
Sep	Oct	Nov	Feb	May
Oct	Nov	Dec	Feb	May
Nov	Dec	Jan	Feb	May
Dec	Jan	Feb	May	Aug

Table 5-3: Expiration Cycles, *con't.*

March Cycle

Expiring Month		Available Months		
Jan	Feb	Mar	Jun	Sep
Feb	MAr	Apr	Jun	Sep
Mar	Apr	May	Jun	Sep
Apr	May	Jun	Sep	Dec
May	Jun	Jul	Sep	Dec
Jun	Jul	Aug	Sep	Dec
Jul	Aug	Sep	Dec	Mar
Aug	Sep	Oct	Dec	Mar
Sep	Oct	Nov	Dec	Mar
Oct	Nov	Dec	Mar	Jun
Nov	Dec	Jan	Mar	Jun
Dec	Jan	Feb	MAr	Jun

USING OPTION TRADING SYMBOLS

An option symbol is composed of three parts: the root symbol, the strike month code, and the strike price code. For example, IBMJD is the option symbol for the IBM October 120 Call. IBM is the root symbol, J is the strike month code, and D is the strike price code. Root symbols for exchange-listed options can be one, two, or three characters long. For over-the-counter stocks, the root symbol is normally three characters ending with the letter Q.

When there is an event such as an uneven stock split, stock dividend, merger, spin-off, etc. that affects the specifications of the underlying contract, the exchange will issue an adjusted option symbol for the outstanding option contracts. This is necessary because one option reflects 100 shares of the underlying stock. The exchange will also issue an adjusted option symbol when an exceptional number of strike prices exist for a specific stock.

The strike month code specifies the expiration month of the option and whether it's a put option or a call option. (For listed exchange options, the actual expiration date is the Saturday following the third Friday of the expiration month. That is the deadline by which brokerage firms must submit exercise notices to the OCC. The exchanges and brokerage firms have rules and procedures regarding deadlines for an option holder to notify his brokerage firm of his intention to exercise. Contact your broker for specific deadlines.)

   The strike price code represents the price per share at which an option may be exercised.
   If the standard strike price codes cannot be used because of adjustments such as uneven stock splits or stock dividends, the exchange will issue an adjusted non-standard strike price and code.
   When using a computer or terminal for quotes you will need the option symbol. It's also handy for you to use these when giving orders over the telephone to your broker. With use, these will become second nature to you.
   The expiration month and strike price codes are listed in Table 5-4.

Table 5-4: Expiration Month and Strike Price

Month	Codes		Call	Strike Price Codes		
January	A	A	5	105	205	305
February	B	B	10	110	210	310
March	C	C	15	115	215	315
April	D	D	20	120	220	320
May	E	E	25	125	225	325
June	F	F	30	130	230	330
July	G	G	35	135	235	335
August	H	H	40	140	240	340
September	I	I	45	145	245	345
October	J	J	50	150	250	350
November	K	K	55	155	255	355
December	L	L	60	160	260	360
		M	65	165	265	365
		N	70	170	270	370
		O	75	175	275	375
		P	80	180	280	380
		R	90	190	290	390
		S	95	195	295	395
		T	100	200	300	400
		U	7½			
		V	12½			
		W	17½			
		X	22½			

The ticker/quotation symbol for the stock is used first, followed by the month and the strike price codes. For example, ABCLW is ABC stock Dec 17 ½ call. ABCFF is ABC Jun 30. XYZHX is XYZ Aug 22 ½.

The option expiration months are the two near-term months plus the two additional months in the January, February, or March quarterly cycle.

THE STANDARD METHOD OF USING OPTIONS

If the option is called (also called assigned or executed), the profit or loss to the option writer is the sum of the cash premium plus the difference, if any, between the strike price and the original stock price. If the stock price rises above the strike price and the stock is called away, the opportunity to profit from further increases in stock price is lost. If the stock price declines, the writer is protected against loss to the extent of the premium.

When the underlying stock stays the same, you win when selling out-of-the-money or at-the-money call options. You keep both the premium and your stock.

When the underlying stock goes up in value, so does the option premium. In the standard or usual manner of option selling, your stock would be assigned.

MY METHOD OF USING COVERED CALLS

All the "experts" in the stock market field will say, "The writer of a covered call option, in return for the cash premium received, foregoes the opportunity to benefit from an increase in the stock price that exceeds the strike price of the call option. The option writer continues to bear the risk of a sharp decline in the price of the stock. The cash premium received will only slightly offset this loss."

This is not correct!

A covered call writer may cancel the obligation any time before it is assigned by executing a closing purchase transaction, that is, buying back the call option that was previously sold—a fungible action. In other words, you may not be buying the option back from the same person who bought your contract, because you have no idea who that is, but you're buying the exact equivalent, so for all practical purposes, you've bought back the one you sold. This is how to make the most of fungibility—to your advantage.

With my covered call strategy you no longer need to care much about the price of the stock you bought. Investors normally watch the prices of their stocks as they go up, down, and sideways. With my method, when the stock does go down, we would buy back the covered call option at a very inexpensive price and immediately write it again.

Perhaps we took in a cash premium of 2 and could close it out by buying the option for 25 cents. If the stock price went down $5 we would write a new covered call at a $5 lower strike price. Since you already lost when the stock declined, using my method you're always taking in additional premium income, which will help offset the decline in the stock price. I call this the parachute effect. Taking in new monies relieves most of the loss in a stock's descent.

When the stock does not reach the strike price, let the covered call expire, keep the premium income, and write a new covered call at the same strike price.

When the stock goes up, you could let the covered call get exercised, at a profit (as in the standard method). Or, with my method, you could buy the covered call back and immediately write a new covered call at a higher strike price, reflecting the gain in the stock price. The second premium added to the first will help defray this cost.

While most covered calls are in existence, both the option price and the stock price will fluctuate. The time value portion of the covered call always represents the judgment of traders. Changes in the time value or the intrinsic value, either of which can affect the covered call price, occur continuously during market trading hours.

For the buyer, the covered call contract is a wasting asset; its value decays as time passes. The time value portion of the premium value is always zero at expiration. Selling the time value repeatedly for the same underlying stock is what makes covered call premium income work for you.

By using the principles in this book, you will learn to react to the stock market. You will not be looking for the stock to go up to make money. You will be making money on the wasting asset called time value.

Your plan is to gain from the time values of the covered calls you have sold. Your philosophy about the stock market will be changed. You will be counting real cash premiums put into your account by speculators.

Once you're moving in the direction of your goals . . . nothing can stop you. Anonymous

CHAPTER 6
MARGIN—THE CREDIT
YOU CAN USE

A bank is a place where they lend you an umbrella in fair weather and ask for it back when it begins to rain. Robert Frost

I have been using margin in my option income portfolio since the crash of 1987. That was the opportunity I had been waiting for: To buy common stocks at up to a 50 percent discount. The money to pay for the purchases came from my use of margin.

Margin is the borrowing of money against the market value of a portfolio. The use of margin enabled me to buy more stocks.

As an investor, you're in business—the business of investing in other businesses through the ownership of common stock. Your shares are your inventory. When you add to your inventory at bargain prices, you will have more shares to sell covered calls against, thus enhancing your earnings. It's a good business practice to build a widely diversified portfolio of good company stocks, and by borrowing against your stock portfolio, you can buy more shares. Again, this enables you to sell more covered calls, thus increasing your profits. Through diversification, you're spreading your risk among many companies, whose stocks may advance or decline at different times. The average of rises and declines over a long period will tend to even out.

Today I owe more in my margin account than I ever have owed in my entire life. However, I still believe in fiscal responsibility and have no other debt.

Your stocks can act as security when you're borrowing money for any purpose through your margin account. Using your margin for purposes other than your investment portfolio, however, could be hazardous to your wealth by jeopardizing your investment plan.

The simple fact is that the use of credit in buying and selling stocks can be no worse or no better than the use of credit in any other business. The skill and judgment of the user of the credit are what is important.

ADVANTAGES OF MARGIN

To use margin, you should understand it thoroughly. The use of margin does generally incur somewhat greater risk and portfolio volatility. Yet, greater rewards are possible with the prudently planned use of margin. Over a long period, as the market goes up and margin is used effectively, margin buying gives you an investment portfolio leverage. In financial terms leverage means using your own money with borrowed money to increase the total rate of return.

The margin loan is open-ended: It has no specific time limits. No specific installments are due nor are principal payments required. The cost of borrowing on margin is very low. Margin loan interest rates are comparable to, if not lower than, the prime interest rates offered to a bank's best business customers. Within certain limits, margin loan interest can be tax-deductible as well.

Applying to use margin is fast and easy. It's part of the application for a brokerage account. There is no credit screening because the stocks in your account provide collateral.

In theory, the margin debt is callable. Carefully read the margin agreement papers you must sign to use margin. But I have never heard of a margin debit called other than to meet the minimum legal requirement.

The cost of borrowing on margin is based on the broker's call rate, a figure published daily in the financial press, plus a percentage added by your broker. The broker's call rate can be found under Money Rates, listed as Call Money. A brokerage agent can tell you the current margin rates.

MARGIN INTEREST

Margin interest is based on the total amount of the margin loan. Keep only one margin account; this will save you interest because the larger the loan, the lower the interest rate. If you have multiple margin accounts, you can transfer them into a single account without having to sell anything. This makes consolidating margin accounts easy.

Margin interest is calculated with the following percentages added to the broker call rate:

Amount of Loan ($)	Broker Add On (%)
0–$9,999	2.00
10,000–24,999	1.50
$25,000–49,999	1.00
$50,000+	0.50

Table 6-1 compares margin rates to other interest rates.

Table 6-1: Margin Rates *vs.* Other Interest Rates (August 1997)

Broker call loan rate	7.25%
$10,000 margin loan	8.00%
Prime rate	8.50%
Credit card rate	18.00%

Your brokerage statement will report your current margin buying power (how much in securities you can buy with available marginable securities), your outstanding margin debt, the interest incurred for the period, and the interest rate for the period. You can get an up-to-the-minute report by calling your broker.

Interest charges begin only upon settlement. Once you decide to use margin, the interest rate, based on the broker's call rate, will be charged each day until you pay off the loan or until you sell the securities used as collateral.

Payback on the loan is at your convenience and there is no fixed payment schedule. Any dividends or interest from the securities used as collateral may be applied to reducing the balance.

A benefit of using margin is that it allows you to get the benefit of the value of your assets without selling them. You don't have to liquidate stocks that are doing well when another attractive investment opportunity comes along. You don't have to realize a profit and pay tax on the sale of stock to use the money.

OPENING A MARGIN ACCOUNT

There is no charge for opening a margin account; margin is simply an additional feature of a brokerage account. If you already have a broker-age account, adding margin requires only that you:

- Read, understand, complete, sign, and return a margin agreement;
- When approved, buy or deposit eligible securities in your account to be used as collateral;
- Instruct your brokerage representative to deposit your securities in a type 2 margin account (type 1 is a cash account); and
- Ask your brokerage representative how much cash or additional buying power is available to you.

IMPORTANT MARGIN PRINCIPLES

Borrow Less than the Full Loan Value: By borrowing less than the full loan value of your securities, you still employ leverage and low-cost bor-rowing but you reduce the chance of having dramatic market fluctuations place you in a margin call (maintenance) situation.

Borrow against Conservative Investments: Borrow only against sound stocks, those with proven track records and dividend payment his-tories; this will lessen the risk of margin call situations considerably.

Borrow against a Diversified Portfolio: It would be highly unlikely that all your stocks would go down substantially at the same time. Some stocks may be lower, some may be higher, and many would retain the same values.

MARGIN TERMINOLOGY

Following are a few margin terms with which you should familiarize yourself.

Market value is the price at which a security is currently trading.

Mark-to-the-Market is the revaluing of a margin account to assure compliance with maintenance requirements. This process reflects daily gains and losses.

Debit balance is the money a margin customer owes a broker.

Credit balance is the money a broker owes a customer.

Equity is market value less the debit balance. Market price fluctua-tions can either increase or decrease market value and equity:

when the amount owed is more than 25 percent of the current value of your margin account. Brokerage firms may set their own higher margin levels.

Your equity (market value – debit balance = equity) must drop to 30 percent before there will be a margin call. Generally, you're given five business days to meet a margin call. With some brokers, the margin call is met if the market value of a portfolio increases in price by the fifth day.

However, with my covered call option strategy we do not anticipate ever receiving a margin call.

A note of caution re-sounded: Your stocks can act as security when you borrow money for any purpose through your margin account, but using your margin for purposes other than your investment portfolio could be hazardous to your wealth by jeopardizing your investment plan.

CONVERTING DIVIDENDS INTO CAPITAL GAINS

Margin interest is deductible against dividend income. Generally, the dividend yields from a total portfolio will tend to offset the margin interest. The use of margin provides the opportunity for you to own a larger stock portfolio. This results in the opportunity for greater capital gains. When you sell the stock, the growth in value is taxed at the favorable long-term capital gains rate, whereas the dividends would have been taxed as ordinary income.

In the next chapter I'll show you how to manage your covered call portfolio.

Money and time are the heaviest burdens of life, and the unhappiest of all mortals are those who have more of either than they know how to use. Samuel Johnson

- Margin interest increases the debit balance, decreasing equity
- Expired covered calls and dividends increase equity, decrea
  the debit balance.

MARGIN AND PORTFOLIO VALUATION

Assume you've opened a margin account with $7,500 cash or margin
securities. If you borrow $2,500 to buy more stock, your portfolio wo
consist of $10,000 worth of stock and would have a $2,500 debit bala
and an equity value of $7,500.

Marginable stocks: market value	$10,000
Debits (loans, margin interest)	−2,500
Equity (net worth)	$7,500

Note: Market Value − Debit Balance = 75% Account Equity

The annual dividends will more than cover the annual margin cos

Marginable stocks - market value	$10,000
Dividend income ($10,000 at 3.5%)	350 annually
Margin interest ($2,500 at 7.0%)	−175 annually
(5% Broker call loan rate	
+ 2% Broker add-on)	

The seasoned margin user spends appreciated gains to buy mo
stock to bring the portfolio to a 25 percent margin level. This presents
more risk than the original decision to use margin.

The decline in the price of one stock will not by itself create a ma
gin call since evaluation of a margin account is based on the enti
portfolio.

Securities valued at less than $5 a share have no loan value. Certai
over-the-counter stocks also cannot be margined. Call your broker fo
details.

MARGIN CALL

The *margin call* is literally a call from your broker asking you to add
assets to your margin account. The Federal Reserve policy requires a cal

CHAPTER 7
MANAGING A
COVERED CALL
INCOME PORTFOLIO

Half of our life is spent trying to find something to do with the time we have rushed through life trying to save. Will Rogers

GOALS AND OBJECTIVES

Let's take a minute here to recapitulate.
The portfolio approach to selling covered calls endeavors to:

- minimize risk;

- provide diversification;

- maximize capital gains potential, dividend income, option premium income, and downside protection;

- earn consistent returns on investment throughout the stock market cycle; and

- increase long-term capital appreciation and income from stock ownership.

The covered call income portfolio is a continuous investment strategy.
Selling covered calls with the strike price near the current market price of the stock usually results in the most balanced combination of potential returns and downside protection. The potential for stock appreciation remains; meanwhile, the substantial covered call premiums that will be earned provide downside protection.

Investors generally have two objections to selling covered calls: They may lose if a stock's price declines past the breakeven point, or calls may limit the upside potential if a stock's price appreciates more than the cash premium.

Limiting an account to one or two covered call positions increases the odds that one of these unfavorable events will occur. A diversified portfolio, however, permits losses on one position to be offset by gains on another.

Assuming the portfolio of stocks selected performs about equally to the stock market, continuous covered call selling should do better than the market. In addition, total returns will be greater over time.

The asset base for a covered call income portfolio should consist only of common stocks for which option contracts are routinely traded on option exchanges.

There are many stocks from which the conservative investor may regularly gross 25 percent a year using covered calls.

The covered call option contract is a wasting asset. Its value decays as time passes. It's the time value component of the covered call contract, not the intrinsic value, that wastes as the expiration date approaches. At expiration the time value is zero.

Time value becomes a money machine as a new covered call replaces the expired call. Because time value sales can be repeated continuously and indefinitely, the profit from time value is a certainty. As one covered call contract expires, a new contract is sold on the same round lot of stock.

Though a covered call income portfolio can be operated with as few as one to five different stocks, for safety it's advisable to increase the number of companies represented in a portfolio, though the number of stocks owned should probably not exceed 20. At this level, diversity and reasonable safety are achieved.

As with all businesses, when initial capital is lower than it ought to be, you make compromises. You accept some higher cost and risk. The next chapter will deal with the problem of insufficient capital.

There are rarely more than 20 stocks that will have call options selling with acceptable time value in their covered call premiums. Most time values are too low. Always be on the lookout for higher time values.

As opinions on the outlook for a specific stock become more favorable, the time value in the covered call premium becomes larger. Sell stocks that no longer have large enough time value premiums compared to the ones you could replace them with, and buy stocks that have larger time values. Follow the changing time values in the financial press.

THE WRITING POSSIBILITIES

So far as strike price and the stock price are concerned, there are three choices:

1. *At-the-money*: The strike price and the stock price are the same. The buyer pays for time value only, as there is no intrinsic value.
2. *In-the-money*: The strike price is below the stock price. The buyer pays for both intrinsic and time value.
3. *Out-of-the-money*: The strike price is above the stock price. The buyer pays less for time value and there is no intrinsic value.

With a little practice you can quickly scan the option page and identify the good covered call writes. Underline each and go on. In a short time, you'll have five to 10 candidates. Then use the *stock and option selection* formula:

Strike price – today's stock price + option premium ÷
today's stock price x sales per year* = annualized percentage

*Using three-month option periods, you could do four sales per year.

The annualized percentage will serve as a realistic guide to selecting the most profitable covered call writes. At this time you may wish to do your fundamental buy-and-hold analysis on each marked stock. Then select the finalist for your covered call income portfolio.

Stocks and their covered calls tend to rise and fall over the weeks and months. At times there may be no stock candidates to study. At other times there will be more candidates than you can possibly buy.

There is never a rush in this business; if you miss one today, there is ample time to find another. This permits you the time to do the necessary investment study on your stock candidates. Everything is constantly changing. The stock market is an alive and breathing business. When you call your broker this morning, the quotes you get will be different from yesterday's closing price. *Don't chase stocks and their covered calls*.

Stock Selection for Covered Call Writing

In shopping for stocks you'll use a different philosophy from the average stock investor who is looking for a stock that will go up in price. Your planned gains arise from the time values of the covered calls you'll sell.

This is an unusual approach to stock selection. Most investors select stocks on either fundamental analysis or technical analysis. You'll use the covered call's time values of a stock, tempered by fundamental analysis and long-term hold principles.

To select a stock for your covered call income portfolio you must have available a current option page. Select a stock from the option page that has the most profitable time value in its call cash premiums. If this stock meets your selection criteria, buy it as an underlying stock.

To get an annual return of 20 to 40 percent you must find available call option cash premiums whose time value will produce a return of 5 to 10 percent in three months on the price of the stock.

Using the option page, you mentally calculate the percentage of the stock purchase price that the time value represents. Of the more than 1,500 optionable stocks, in all probability, you'll only have some five to 10 stocks to consider. If the time value seems attractive, then turn to fundamental analysis to make your decision.

There are approximately 30 investment strategies using stocks and options. You'll be using only one. You'll be writing covered call option contracts. At expiration, you'll sell a new covered call contract for the next 90-day maturity. This investment strategy attempts to get the maximum gross profit while keeping expenses down, thus generating as great a net profit as possible. The main expenses will be stock brokerage commissions.

Brokerage Commissions

Commissions for covered call trading are less than for the purchase and sale of common stocks. Keep stock turnover at a minimum. Sell stocks only when there is a real reason (e.g., the time value of the premium is smaller than can be had with another stock).

Commission expense with covered calls, as with stocks, is less per trade when you're dealing with more volume or value. There are savings on commissions when you sell multiple contracts on covered calls. Five contracts do not cost much more than one. There is an economy in scale in trading covered calls.

THE BUY/WRITE STRATEGY

Make your time value comparison of the optionable stocks you own. If a stock rates poorly, sell it and buy a promising one. Otherwise, immediately sell a covered call to protect yourself against a price decline and to generate current income.

When buying an optionable stock, immediately protect yourself against a price decline. *Buy/write* is the investment strategy of purchasing stock and writing covered calls simultaneously. This is a conservative approach to generate maximum current income by using the call option cash premiums. As an example: Tell your broker to buy 200 shares of ABC common stock and sell (write) two contracts of the ABC September 10 call covered calls with a net debit of $9.00.

By doing a buy/write order, if the stock price was $10 and the three-month call was $1, the amount owed would be the difference of $9 per share or $1,800 plus commissions.

Option Premium = Intrinsic Value + Time Value

Buy ABC for $10.00 a share	−$10.00
Sell an ABC 10 option for 90 days	+1.00
Out of pocket	−$9.00
Total agreed price and premium	$11.00

You made 10 percent in 90 days (40 percent annualized). You can figure the desirability of stock/call option choices from time values alone, as derived from the covered call cash premium, stock cost, and strike price. You do not need to add the dividends, commissions, and margin interest, as this often complicates a simple procedure. Remember, we're selling time—when a covered call option contract expires, the contract is void forever. The call option maturity months are integral to the quotations.

A typical quotation for a specific call option appears below. At a strike price of $20, this option is in-the-money by 7/8 ($0.875).

Date: April 21 (April options have expired)

	Jul	Oct	Jan	NY Close
ABC Corp. 20	2 7/8	4 3/8	5 1/2	20 7/8

Intrinsic value is $0.875. Time value for July $2, for Oct $3.50, and Jan is $4.625. The stock closed that day at $20.875. If you sold one ABC Corp. July 20 covered call contract at $2.875 per share for the 100-share contract, you would receive $287.50 gross income. You would receive $437.50 for October and $550.00 for the January covered call contract.

It's advantageous to do three-month time frames on covered call contracts, instead of six- and nine-month contracts. The premium money looks larger at first for doing a nine-month contract but note the results below, if the time value remained the same. Consider the following:

	3-Mo.	6 Mo.	9-Mo.
First contract	$287.50	$437.50	$550.00
Second contract	$287.50		
Third contract	$287.50	$218.75*	_____
Gross	$862.50	$656.25	$550.00

*Half of the second 6-month option expires next year. Thus, only $656.25 is realized this year.

You'll realize $206.50 more ($862.50 – $656.25) and $312.50 more ($862.50 – $550.00 = $312.50) by writing consecutive three-month covered call contracts. Other advantages will be discussed later.

On buying a new stock and selling a short-term covered call, if you're mid-cycle for the 90-day contract, do not sell a call for the next contract expiration date but go to the second contract expiration.

FORMULA FOR STOCK AND OPTION SELECTION

If an option is written at-the-money or out-of-the money, the call option premium is all time value. The in-the-money call option will have time value plus the intrinsic value above the strike price.

The formula for computing estimated *annualized* rates of return in percentages is

Strike price – today's stock price + option premium ÷ today's stock price x sales per year* = annualized percentage

*During a year you can write four three-month options, two six-month options, and one and a quarter nine-month options.

Using the April 21 example for the July, October, and January options, the percentages are as follows:

Strk. pr.	Today's − stk. pr.	+	Op. prem.	Today's ÷ stk. pr.		Sls/ × yr.	= Annualized %

June 20

20 − 20.875 = −0.875 + 2.875 ÷ 20.875 = 0.0958 x 4   =   0.3832 or 38%

October 20

20 − 20.875 = −0.875 + 4.375 ÷ 20.875 = 0.1676 x 2   =   0.3353 or 34%

January 20

20 − 20.875 = −0.875 + 5.50 ÷ 20.875 = 0.2215 x 1.25 =   0.2796 or 28%

This formula will let you do your percentage calculations rapidly.

Within the last two or three days of an expiring covered call, if the stock price is below the strike price or even with it, the covered call option will expire worthlessly. If your stock price is above the strike price (in-the-money) at expiration, you can either let them have it (standard method), or you can buy the fungible covered call to close before expiration, keep the stock, and write it again at a higher strike price.

After the expiration in July, assuming we wrote the covered call in April, we would look at the October expiration. The October expiration is the next one available on the three-month cycle after July. You must learn to react *before* the option expires.

We have a method that guarantees a steady income by selling time only and does not try to make profits in the irrational market. We are no longer stock pickers. We are not trying to capture those elusive stock price swings by being a market timer. We are just cashing in on the decaying time values we are selling.

THE CALL OPTION BUYER

After many years of selling covered call options, I still marvel that such an opportunity exists when I see the option monies come in. I have the option buyers to thank for making this possible. There are more option buyers than there are option sellers, which helps keep option premiums up.

You must understand that option buyers are speculating. They plan for the stock price to rise sharply beyond the value of the cash premium they paid to you, betting that the call option contract can be sold at a profit before expiration without their having to buy or call the underlying stock.

The covered call option contract that you sold once may be traded dozens of times. You'll not care about this. The buyers are gambling with small amounts of money, and do not have the cash to buy your stock from you. They do not want the stock. They want the rapid leveraged gains that can occasionally be made.

A COVERED CALL INCOME PORTFOLIO =
A PROVEN WINNER

With call options we have a win-win-win situation; with stocks alone, a win-lose-draw.

Underlying Stock Price	Declines	Increases	Same
Stock with Call Options	Win	Win	Win
Stock Only	Lose	Win	Same

If the stock price increases, you keep the time value portion of the cash premium received, even if the option holder exercises the right to buy your stock. The intrinsic portion of the call option goes to the buyer of the call option. Though it may appear that you give up the gain in a large price rise that you would have had if you had not sold a call option, you'll use it to your advantage. In a small price rise, if the cash premium received is larger than the rise in the stock price, your gain will be larger than the gain in the stock price.

If the stock price stays the same, the covered call will expire and you keep the cash premium received.

In a price decline, if the covered call cash premium received is larger than the decline, you have no loss, and may still have a gain. The only risk is when the stock price goes lower than the cost of your underlying stock and the cash premium received. It's precisely at this time that you should buy back the call option to close, for pennies on the dollar, and immediately write a new covered call option. Because you'll always be taking in money, the income will act like a parachute in a stock price decline.

The stock market is irrational and any stock price has an equal probability of going up or down. This will protect you. You have reduced, though not eliminated, the possibility of a loss. By using the guidelines outlined in this book, you can react to the market.

You have just learned how to protect your covered call income portfolio from a decline in stock market price value while you continue to sell covered calls.

COVERED CALLS IN RETIREMENT ACCOUNTS

A covered call income portfolio meets all the objectives of retirement accounts, such as SEPs (Simplified Employee Pension Plans), IRAs

(Individual Retirement Accounts), Keoghs, defined contribution plans, and defined benefit plans. Retirement accounts should conserve, capture, and grow capital to financially prepare individuals for their retirement years. Establish retirement accounts as early as possible.

There is nothing about investment policy in either the ERISA (Employee Retirement Income Security Act) or the IRA act that restricts any particular strategies. However, the objective of these accounts is protection and growth. Which strategies are allowed is a function of the tax code guidelines from the Internal Revenue Service (IRS) and the brokerage firm's policy or the Master Trust Agreement.

All retirement plans covered by ERISA allow only the strategies mentioned in the Master Trust Agreement. This obviously will tell you which strategies are eligible in your particular plan. Strategies that require the account to use margin (borrow money) are not allowed, but any strategy that uses fully paid stock or is cash secured may be acceptable. Check with your IRA trustee, broker, or tax professional about the types of trading allowed in a tax-deferred account.

Even if the retirement account allows these strategies, the management firm must approve your use of them. All covered call investing, whether in a retirement or a non-retirement account, is subject to the consideration of customer suitability.

Now that you understand why option strategies may or may not be permitted in a retirement account, let's look at the covered call income portfolio strategies that are generally permitted and how they can help you in planning for your retirement.

The opening strategy of a covered call income portfolio is the covered write. The covered write consists of selling a covered call against stock currently held. The seller receives a cash premium in exchange for being required to sell the stock at a set price (the strike or exercise price). This is a very common strategy in an IRA, retirement or non-retirement, account. For example:

You own 500 shares of RST at $43.
You sell five RST six-month 45 calls at $3.50.

You'll receive $350 ($3.50 premium x 100) per contract in exchange for being obligated to sell the stock at $45, no matter how high it should rise for the life of the contract. You have lowered your breakeven on the stock to $39.50 ($43 − $3.50 premium received) and limited your sale price to $48.50 ($45 strike + $3.50 premium received). Covered call writing makes particularly good sense in retirement plans because the premiums received for selling covered calls compound tax-free.

If the stock is at $45 or higher at expiration in six months, sell the stock for the strike of $45. You'll have received $45 for the stock and $3.50 per share for the covered call, making $48.50, which is an annualized return of 25 percent on the $43 stock. You can use this money to do a buy/write on the next stock for your covered call income portfolio.

Of course, if the stock price is not at the strike price on expiration six months from now, you keep the premium and you're free to do it again. The premium you kept is $3.50—a 16 percent annualized return on a flat stock. Not bad!

If the stock drops in price rapidly to below $40, follow-up action is necessary. It's advisable to buy back your covered call if the price is one-half or less of the cash premium received, and sell a new covered call for the next full option expiration cycle. If it's midway (45 days), and you can buy close to the covered call for $1.75, you'll retain $1.75 of the time value. By selling the new covered call to open you'll bring in more new money.

FOLLOW-UP ACTION

Once a covered call is sold it must be monitored, since follow-up action must be taken at or before the expiration of the call option—even if the action is just a decision to allow it to expire.

Some investors prefer a passive approach. They allow the stock to be called if its price is above the strike price at expiration. They rewrite a covered call if the stock price is below the strike price at expiration. This approach is simple and functional, but more active management creates greater profits.

Follow-up action on a covered call is guided by movements in the price of the underlying stock and by the passage of time. Consideration is also given to the stock price in relation to the strike price of the covered call sold.

Covered call selling is a strategy designed to provide a balance of returns consisting of potential for stock appreciation, income, and downside protection. As time passes or the underlying stock fluctuates in price, this balance will change. Once the balance changes, it's time to consider action to restore the position to its original balance, or liquidate it.

The man on top of the mountain didn't fall there. Anonymous

PERIODIC REVIEW

Buy more stocks? Sell the stock? Hold the stock? Informed investors agree that periodic review of their portfolio equity holdings is part of the investment process. In the most simple form of review, the investor looks at each equity holding and asks whether or not it should still be held. Are the fundamental and other reasons this stock was purchased still in effect? Should it be sold? Has the stock met a set objective or changed to the point where holding it can no longer be justified?

A more elaborate review process would add an additional question: Should we be adding to our current holdings because the stock has moved down to an attractive buying range? This simple review process can be summarized by questioning whether a holding should be bought (adding to positions), held (doing nothing), or sold (liquidating).

Some investors might answer as follows:

1. Would I add to this position?   Yes, but at a lower price.
2. Would I sell this stock?   No, hold the stock.
3. Would I sell this stock?   Yes, but at a higher price.

Let's refine our review process and ask our questions as follows:

1. Would I be willing to add to this position if my costs were 10 percent below the current market price?
2. Would I be willing to liquidate my stock at a price 10 percent above the current market price?

Investors who answer yes to both questions can either wait for the stock to move up or down 10 percent before taking action—or they can take immediate action, use covered calls, and create an opportunity to increase the return of their holdings, even if the selling or buying target is not realized.

Here is how it works. Investors who are willing to sell a holding at a higher price can write covered calls against their holdings. At expiration, if the stock price exceeds the strike price, the stock is sold.

Investors wanting to add new stock could write covered calls at the stock price or lower. The cash premium received could be thought of as getting a discount on the stock.

For investors who are willing to sell their holdings at a higher price and add to their positions at a lower price, this is a key strategy to consider. Table 7-1 shows a hypothetical portfolio and options. For simplification, taxes and commissions have not been factored in.

TABLE 7-1: October Calendar Date—January Call Quotes

Stock Symbol	Stock Price	Series	Price
ABC	43 $1/4$	Jan 45	2 $5/8$
LMN	90 $1/2$	Jan 95	2 $13/16$
XYZ	23 $1/4$	Jan 25	1

Now assume the investor who owns the portfolio in the table is willing to sell any of the holdings at 10 percent above the current market price. He would also be willing to double-up if the price were 10 percent below the current market price.

With ABC stock at $43 $1/4$, an effective selling price of $47 $5/8$ and a purchase price of $38 $7/8$ are required to meet these objectives.

Should the investor sell the stock on option expiration at $45, he will get to keep the premium. Thus the effective selling price will be $47 $5/8$, which was the target.

If the price of ABC fell to $41 $5/8$, the investor could buy more at that price and with the premium received from optioning the new stock reach the target of $38 $7/8$.

TABLE 7-2: January Expiration—Required Prices to Reach Target

Stock Symbol	Stock Price	Sell Price	Buy Price
ABC	43 $1/4$	47 $5/8$	40 $5/8$
LMN	90 $1/2$	97 $13/16$	87 $5/8$
XYZ	23 $1/4$	26.00	22 $1/4$

As you can see in Table 7-2, the 10 percent above/10 percent below objectives can easily be met. With covered calls, investors can increase the return of their holdings when neither the buying nor selling targets are met. Look again at ABC stock. If at the January expiration the stock is trading at the same price, it's highly unlikely that the call option will be assigned. It will expire worthlessly. In this case the investor keeps the $2 $5/8$ per share and repeats the operation, selling the April 45 calls.

The returns for our stock portfolio in a flat market can be seen in Table 7-3. They would be enhanced further by any dividends received on the stocks held.

TABLE 7-3: January Expiration—Annualized Returns in a Flat Market

Stock Symbol	Stock Price	Total Premium	Total %	Annualized %
ABC	43 $1/4$	2 $5/8$	6.0%	24.0%
LMN	90 $1/2$	2 $13/16$	3.2%	12.8%
XYZ	23 $1/4$	1	3.3%	13.2%

As illustrated, the mechanics of covered calls are quite simple, and the strategy offers excellent returns in both flat and rising markets, while letting the investor average down in falling markets.

The following points should be taken into account before establishing a position.

1. *Up/down targets*: There is no magic to the 10 percent targets selected in these examples. They are simply realistic expectations for stocks with average volatility. Wider targets can be established for more volatile stocks whose covered call cash premiums are normally higher. Narrower targets should be considered for lower volatility stocks. Targets may be established for longer or shorter call option time periods.

2. *Time horizon*: A three- to five-month time period will let investors set targets that meet realistic expectations. This may provide an adequate return in flat markets. Before establishing the strategy, ask yourself: "Would I be willing to sell/buy my stock 10 percent above or below the current market price during the next three to five months?"

3. *Future stock value*: In our examples, when a call was sold, strike prices were close to the current stock price. You'll probably sell a call at a higher strike price if you're bullish on the stock and sell at a lower strike price if you're bearish on the stock.

4. *New positions*: Covered calls need not be limited to stocks already in the portfolio. An investor can use a buy/write to simultaneously purchase a stock and sell covered calls against these shares.

Follow-up action mainly consists of monitoring the prices of the stock and the covered call option, taking no action if the underlying stock and its option price go up or remain the same until option expiration week. It's at the option expiration week that the decision is made to do nothing and have your stock assigned, or buy the option back, keeping your stock (closing out the option). If the underlying stock and its call option price go down, wait until expiration and let it expire worthlessly. A better strategy is to buy it back early in the option time period, canceling the contract, and rewrite a new, lower strike price covered call for the next expiration cycle.

How can we determine when to do this buy back? The following questions must be considered:

1. Is the time premium remaining on the call option less than 25 percent of the time premium received? If so, consider writing a new covered call, as most of the profit from this position has been made.

2. Is the price of the call option less than 25 percent of the premium received? If so, consider writing a new covered call, as minimal downside protection remains.

3. Is the stock about to go ex-dividend? If so, it may be necessary to buy back the covered call to protect receipt of the dividend.

COVERED CALL INCOME PORTFOLIO SUMMARY

Do not *wish* (*or pray*) for a profitable trade. Always make trading decisions based on sound analysis.

Ideally, you should own several common stocks in different industries. Diversify, diversify, diversify! This cannot be stressed enough!

Do not let others influence your trading decisions. Make your own decisions, and stick with them.

Try to choose stocks with call options that expire in different months. Use limit orders in your operation.

Let dividend income pay the margin interest costs and covered call cash premiums to reduce the margin debit balance. Then borrow more on margin to buy more stocks and to write more call options.

Do not trade just to trade. Many people enjoy trading for the excitement. Maintain your present position when there are no definite trading opportunities. Be patient and disciplined; many opportunities will appear.

A scissors grinder is the only person whose business is good when things are dull. Anonymous

If thou wouldst keep money, save money; if thou wouldst reap money, sow money. Thomas Fuller

CHAPTER 8
TIMING AND TAXES

The ladder of success doesn't care who climbs it. Frank Tyger

Let us assume that on February 21 you bought 100 shares of ABC Corp., an optionable stock, for $35 a share at a total cost of $3,500. On the same day you sold an opening covered call for three months to expire on May 20 at a strike price of $30, for which you received $850. When you did that, your out-of-pocket cost dropped immediately to $2,650: $3,500 − $850 = $2,650. So, because of the covered call hedge, you're protected in a price decline until the stock price drops below $26.50 per share.

You know that when you buy a stock, the price can go up, down, or stay the same. You also know that for the three-month period of the option, price movement in the stock, and the covered call contract you sold will be related.

Table 8-1 shows what can happen to the market price of the call option value at expiration due to stock price movement.

Table 8-1: Call Option Value at Expiration and Stock Price Movement

	Price Feb. 21	Prices May 20 (expiration day)				
ABC Corp. Stock	35	25	30	35	40	45
ABC Corp. May $30 Option:						
Intrinsic value	5	0	0	5	10	15
Time value	3.50	0	0	0	0	0
Option value	8.50	0	0	5	10	15

Table 8-1 reflects five possible market prices for the stock on May 20th. For each stock price you can see the related prices of the covered call contract. There is no time value remaining. The intrinsic value is the difference between the stock price and the strike price. It's zero if the stock is selling at or below the exercise price of $30.

If the covered call is exercised (called away), you would receive $3,000 for your shares. You already received $850 for the call option, so that makes total cash of $3,000 + $850 = $3,850. You paid $3,500 in stock cost. Your gross profit would be $3,850 – $3,500 = $350, not allowing for stock brokerage commissions; in other words, you would have the $350 gross profit plus your original $3,500.

Your stock will not be called if the stock price is below the exercise price of $30. Stocks are rarely called during the life of a contract. Covered calls are exercised the last few days of the call option, when the time value component of the option premium is very small.

NO TIME = NO TIME VALUE

In our example, the time value on February 21 is $3.50 per share. At expiration, the time value component of a covered call premium is always zero. Understanding this fact permits you to make money in your covered call income portfolio. You always realize the time value as gross profit.

The cash received for the sale, less the commission, will be credited to your brokerage account on the next trading day. You get your money in one day, and if you just bought the stock with a buy/write, you'll have three days to pay for it. Thus, you'll be using other people's money to buy it, because you can apply the proceeds of the covered call sale toward the price of the stock you just bought.

The net cash you receive from this opening sell transaction increases the cash amount carried in your account. When you sell a covered call contract, you cannot determine the tax consequences until the call option has expired in one of three ways:

1. *Exercise*—the buyer of the call option calls the stock away, i.e., you have to sell.

2. *Expiration*—the passage of time makes the call option worthless.

3. *Buy back*—the covered call is bought back with a *closing buy*, eliminating the obligation to deliver or sell the stock.

There are three ways to create a taxable event:

1. *Exercise*—you cannot control.
2. *Expiration*—you cannot control.
3. *Buy back*—buying back your covered call contract is something you certainly can control.

Most financial professionals promote waiting until expiration if the call option is not exercised, totally ignoring the buy back option.

Consider the example of the three-month ABC May 30 call option. On May 20th the stock was under the $30 strike price and no option holder would force you to sell the stock. Clearly the same shares could be had for less on the open market.

At expiration, what actions would you take if ABC Corp. stock was at one of five closing prices: 25, 30, 35, 40, and 45?

If the stock price is at or below the strike price of 30, you take no action. The option contract will expire worthless. This expiration is a taxable event.

If the stock price at expiration is above the strike price of 30, and if you wish to keep the stock, buy an offsetting call contract with a closing buy transaction. Do this just before expiration, at the prices illustrated in Table 8-1. On May 20, with the stock price at $35, you would pay $35 – $30 = $5 to buy the call option back. At $40 you would pay $40 – $30 = $10. And at $45 you would pay $45 – $30 = $15.

To take these actions, instruct your stockbroker to "buy one ABC Corp. May 30 call option to close at the market."

Any one of these closing buy transactions is a taxable event.

TAXABLE EVENTS AFTER THE OPTION

To explain the tax implications further, let's look at the three things a stock price can do after you sell a covered call against it: go down, go up, or stay the same.

Stock Price Goes Down

Let's consider a worst case scenario: ABC Corp. stock declines from your purchase price of $35 (February 21) to $25 a share on May 20. (The very worst case would be if the company went broke and its stock fell to zero.

That's one reason to diversify your portfolio: to minimize the risk of a bankrupt stock.)

In this example the tax consequences are not good. There's a realized gain of $8.50, the option premium, which is taxable as ordinary income. The unrealized capital loss in the stock of $10 a share is a reduction in your equity. The only comfort is the pre-tax hedge given to you by the option premium of $8.50. In spite of the tax negative, this cash flow reduced your pre-tax loss in equity to $1.50 when the stock value went down $10.

This stock should be only one of about 20 stocks you should own in your covered call income portfolio. You would expect to have some short-term losses from other call options you sold, then bought back at higher prices. These net realized capital losses can be applied against your $8.50 net capital gain.

Clearly, buying stock at $35 and selling covered calls for $8.50 is a more conservative way to own stock than if you simply buy stock and wait for its market price to rise. On a pre-tax cash basis, your invested capital would not be reduced until the stock price declined from $35 (your cost basis) to $26.50, a 24 percent drop.

Stock Price Goes Up

No matter how high the stock price may rise, you agreed to sell your stock for $30. But actually you're selling for the $30 strike price plus the $8.50 covered call option premium.

Let's consider the net effect of buying a covered call offset when the stock price is above the strike price (when the call option is trading in-the-money). When you buy the covered call offset, your obligation to sell stock at the exercise price has been canceled. Using the new, higher market value of your stock, you can sell a covered call with a higher strike price.

If the stock price on May 20 is $45, you'll pay $15 for the covered call buy-back (call option offset). This produces a realized short-term capital loss of $6.50 ($15 − $8.50 = $6.50). Your unrealized capital gain in the underlying stock is $10 ($35 up to $45), so you have an unrealized gain in equity of $10 and a realized loss of $6.50.

What you have done is shift assets from one position to another. You picked up an equity gain that is not yet taxable and realized a tax loss benefit that can be used immediately.

In rising markets you'll generate year-to-year tax loss carryovers, normally short term. Any unused capital loss remaining after taking the maximum deduction allowable against ordinary income can be carried over indefinitely until used. Many investors build a loss carryover account, which allows them to realize tax-free cash in the future. Gains realized on a future trade can be offset by losses in the carryover account.

This loss carryover account is valuable because it comes from your covered call buy back activity, which can produce non-taxable gains in your portfolio. You actually gain equity while receiving a short-term tax loss. You make money while generating a tax deduction!

If a covered call is exercised, it may be to your advantage. If you wish to sell the underlying stock at expiration, you simply do nothing. When it's in-the-money it will be called. This is a welcome exercise that you control.

I sell a covered call deep in-the-money knowing that it will be called. It usually does the trick. If the stock should drop below the strike price after writing it deep in-the-money, you can keep a large premium and do it again.

The exercise is of value because of the favorable tax treatment of capital gains. Normally, call option premiums are short-term gains. When the underlying stock is called away, the call option premium may be treated as short or long term depending on how long you held the stock.

The adjusted sales basis of the stock called away is the strike price plus the call option cash premium. If the stock is a long-term holding, the covered call premium will be considered a long-term holding. This gives you the opportunity to turn a short-term capital gain (the call option premium) into a long-term one for tax purposes.

Another exercise may not be so welcome, and it can happen suddenly. You receive an unexpected, unplanned demand from the holder of the call option to exercise his right to buy your stock at the strike price. You do not have any control of the unwelcome assignment. On the same day you're called or assigned, you must comply with the terms of the call option contract: You must sell the stock.

Notice I said you *must comply* with the call option holder's request and deliver stock. It does not have to be your own shares—just the same number of shares for the same strike price. Yes, here is the time to contemplate "fungible" and "fungibility." You can let the holder have your shares—or you can buy shares on the open market and deliver them. Let me show you why you might want to do that.

Whether the exercise against you is welcome or not, it's essential that you understand that in responding to an exercise you do not have to sell the original covered call optioned shares. You can:

1. Deliver shares you already own, or

2. Buy and deliver new shares purchased on the open market at the prevailing price. Selling the new shares to the call option-holder at the strike price fulfills the terms of the contract, using fungibility.

This will be easier to understand if we go back to the last trades we did with ABC Corp. On February 21 we bought 100 shares of ABC Corp. at $35 and on the same day we sold a call option (ABC Corp. May 30 for $8.50). On May 20 the ABC Corp. stock is selling for $45 and your broker informs you that your 100 shares of ABC Corp. stock were exercised for $30.

The adjusted sales basis of the stock called away is the strike price plus the call option premium. If the stock is a long-term holding, the option premium will be considered a long-term holding. This gives you the opportunity to turn a short-term capital gain (the option premium) into a long-term holding for tax purposes.

You can decide to retain your shares. For whatever reason, you do not want to sell them. Maybe you want to hold them long enough to realize a long-term capital gain. Or the stock has gone up from $35 to $45 a share and you would prefer to keep the lower-cost shares in your portfolio to avoid a taxable event. You bought the shares at $35. You wrote the call for $8.50. Now you have been assigned at $30. The tax implications: –$35 + $8.50 + $30 = +$3.50 x 100 = $350.00 short-term taxable capital gain.

You can decide to buy new shares today at $45 and sell these for $30 in cash. You wrote the covered call for $8.50 to sell stock for $30 and buy stock to deliver at $45. The tax implications: +$8.50 + $30 – $45 = –$6.50 x 100 = –$650 short-term capital loss.

Cost basis of 100 ABC Corp. at $35	$3,500
Market value of 100 ABC Corp. at $45	4,500
Gain in stock value (non-taxable)	1,000
Short-term capital loss (deductible)	–650
Net after-tax gain	350

You received a net non-taxable gain of $350, 10 percent in 90 days on your equity of $3,500. Plus your short-term capital loss can be used at tax time, so your annualized, after-tax return on your applied equity is 40 percent. Note the gain in equity that resulted from the time value of the cash option premium at the time of your opening sell.

There has been no decrease in your equity by these trades because you owned the underlying stock. Your after-tax equity increased. As part of the transactions, you obtained a realized short-term capital loss. These losses are not reductions in your equity.

Losses from covered call buy backs may be accumulated during the tax year. If you don't use all the losses in one year, you may carry unused losses into future tax years. Such covered call buy-back losses comprise the tax-loss carryover account discussed previously. We're realizing tax losses without equity losses.

The second decision was to buy 100 shares of ABC Corp. at $45 or $4,500 cash outlay. You sold 100 shares of ABC Corp. at $30 or $3,000. This equates to –$15 a share (–$45 + $30 = –$15); you paid out $1,500 in cash. Your cost basis of the stock is $45 and your sale basis of the stock is $30, for a net loss of $1,500.

Your decision was to keep your original stock and buy and sell the new stock on the same day. You have a gain and tax situation that are the same as though you had used the covered call option buy back offset. Common stock commissions are somewhat larger than option commissions. Thus, the unwelcome call option assignment is slightly more costly than the covered call option buy back offset.

Buying new shares when you're exercised (either welcome or unwelcome) gives you the opportunity of selling the higher-price shares to the covered call option holder. In this example the new shares were higher, so we sold the new shares and kept our lower cost basis. If our cost basis in the original shares was higher, we would sell the older, higher-cost shares and retain the newer, lower-cost shares in our portfolio.

If you sell covered calls on shares of stock that you have held for years and your cost basis is very low, you can always substitute newly acquired shares to comply with the terms of the call option contract. You never have to sell your original, low-basis shares in response to an unwelcome assignment.

Since earlier you had declined decision one, now you must give an order to your broker: "Buy 100 ABC Corp. at the market and sell those shares just purchased at $30 to satisfy the assignment I have received."

Now that you've satisfied the covered call, the underlying stock in your possession is available for writing covered calls again. Any stock that went up 28 percent in three months has caught the eye of the speculators. If you wrote the next period (August calls) and wrote it in-the-money as before, you would get a rich premium. The ABC Corp. August 40 probably would bring in $5 for intrinsic value and $4.50 for time value, so you probably would get $9.50, or $950. We now have a fresh $950, which more than offsets our previous $650 loss.

Stock Price Stays the Same

If the stock price on May 20 was $35, what you paid, you would pay $5 to buy the covered call offset. You would then have a net realized gain of $3.50, taxable as ordinary income, the difference between the $5 payment and the $8.50 premium. You would have no change in unrealized gain for the stock itself. Let's see how you did when the stock price remains unchanged.

Strike Price	Today's Stock Price		Option Premium Price	Today's Stock Price		Options Per Year		Annualized %
$30	$35	+	$8.50	$35	x	4	=	40%

With the stock price unchanged we made 10 percent in 3 months and 40 percent annualized, before commissions. Not bad!

ADJUSTING COVERED CALL WRITES

The most exacting thing to do properly in a continuing covered call writing program is adjusting the call options. A bad adjustment can turn a good covered call write into a calamity. Anyone who has used covered call writes for any period of time has probably realized the probability of adjusting numerous times. It's a serious choice. It's also an essential decision.

Adjusting is changing a position in response to time erosion and changes in the underlying stock price. Adjusting involves the repurchase of a previously written covered call and the sale of a new call option as a replacement. Most often the reasons for adjusting are that the previously written covered call is now in-the-money, it's low on time premium, and the covered call writer wants to avoid assignment, avoid a taxable event, and expand profit potential.

The ideal setting for a covered call writer is a slowly advancing market. Ideally, the underlying stock would move up just to the area of the call strike price. That would produce maximum profit on the original position—and great rewriting prospects. The challenge starts when the stock makes a big move one way or the other.

Our aim here is to reveal the situations that develop when the underlying stock for a covered call has a big gain. For many unaware covered call option writers, this situation doesn't present much of a problem; they just let the stock get called away. After all, the covered call write will have returned its maximum profit potential. Many others will sense that they have missed out on extra profit possibilities, so they keep their stock by buying back the outstanding call options, and they will want to adjust the next calls up. This is where more important challenges could develop.

The advantage that covered call writing offers is time premium income. The method is to generate enough premium income to offset the disadvantages of limited profit potential. Because adjusting up to a higher strike price usually involves a negative cash flow, it may actually seem contrary to the investor's initial objective. This problem can be solved by writing long-life contracts that can help in three ways: (1) The extra-large time premium in the longest contract is the most practical to sell. (2) Going to the longest-term contracts will greatly increase flexibility in making adjustments in the near future. (3) A possible taxable event, such as selling a low-cost basis stock, is avoided.

In determining whether or not to adjust and what to adjust to, there are two rules that can help keep things in perspective.

Rule #1: In any covered call write, the optimum call to sell is the strike price that the stock will be at (or near) on the call option expiration date. This is a tongue-in-cheek way of saying that the stock's market price, not the call option cash premium, is the key to performance in covered call writes.

Rule #2: When adjusting, don't pay to adjust up to an at-the-money call if you don't like the stock enough to buy more at the current price. The critical error in adjusting up is to pay too much to adjust to too high a strike price, only to see the stock drop back down. To prevent this, adjust your strike to midway between where your closed strike was and where the stock is presently. For example, if your closed-out position was a $20 strike price and the market price today is $25, sell a $22.50 strike. This will enable you to keep $2.50 intrinsic value plus the time value of your covered call, which will help with the cost of the buy-back.

With all that in mind, let's go through the alternate views for each of several possibilities. Assume that we bought stock and sold covered calls three months ago. The stock has since gained and is now trading:

- Just Above the Strike Price

In this case we have close to a perfect covered write. We could do nothing and let the stock get called away; after all we haven't missed out on much profit potential. On the other hand, we should have excellent prospects for adjusting.

First check the net credit for adjusting out to three months to the same strike price—in most cases, this will be the preferred adjustment. If the net credit for the trades is acceptable, do them. If not, let the stock be called away.

An acceptable net credit is usually in the area of 10 to 20 percent. Adjusting up to a higher strike and six months out would only be advised if we were still fairly bullish on the stock. Paying to adjust up could actually hurt if the stock is lower at the next expiration. In that case, we would have gone from a nearly perfect position, having paid back some of the original covered call's premium to adjust to a higher strike and will now watch the trade turn to a loser.

Considering the penalty for being wrong in adjusting up in the first scenario, it's advisable to move up to at most a strike price midway between where you were and the market price of the underlying stock today.

- At 10 to 15 Percent above the Strike Price

In this case we're comfortable in our profit position. The stock has performed better than expected and though we missed out on some good profit potential, the advantage is more in favor of keeping the stock than just letting the stock be assigned.

Ironically, it's just this sort of situation that convinces more and more covered call writers to adjust up. The net credit available for adjusting out six to nine months is almost enough to make it worthwhile. Paying to adjust up two or three strike prices provides an attractive ratio of cost to profit potential; unfortunately, it increases the chance of, and the penalty for, being wrong. Assuming we're still bullish enough on the underlying stock, adjusting to the first in-the-money strike is a reasonable decision. If that adjustment is too expensive in terms of cost per dollar of new profit potential, let the stock go and look for new opportunities.

- Way Above the Strike Price

Don't let it go! In such a case, you'll probably wish you had never written the call in the first place. Don't let your emotions lead you to make an even greater mistake by not adjusting up. Adjusting to an in-the-money is probably too expensive and inefficient. Going to a longer-term contract with a larger premium and a midway strike price produces the most cost efficiency. If the adjustment is too expensive in terms of cost per dollar of new profit potential, let the stock go and look for new opportunities.

The purpose here is not to suggest that covered call writers should never let every in-the-money covered write be assigned. Just be careful. Adjusting introduces a misleading new type of risk to a covered call write. Be sure of what you're doing, know the risks, and don't take chances on a possible major blunder.

Rule of thumb: Buy back your covered call if the price is no more than 50 percent of the premium received, and sell a new covered call option for the next full call-option expiration cycle. If it were midway (45 days), you could buy back the call option for $1.75, resulting in your retaining $1.75 of the time value as well as the $5.00 intrinsic value. By selling the new covered call further out, you bring in more new money.

ONE STOCK FOR ONE YEAR

To obtain the best results, all stocks should be working for us at all times. Therefore, when you close a covered call on a block of stock, immediately open another using a call spread. This works like a buy/write, where two orders have to be filled simultaneously: buy to close call options and sell to open them for a net credit or net debit.

Table 8-2 shows a simulation of potential results for XYZ Co. The purpose is to show the types of results falling stock prices have and what can be done.

Simultaneously, using a spread order, we wrote the July 17 ½ options at 1 ¾ to net $875. In May we bought these back at 1 ¾ for a profit of $500.

We continued to sell and buy back until November. Then, anxious to set up a tax loss, we bought back the January 15 options at a cost of $4, or $2,000, for a short-term loss of $500 to be applied against other income. The same day we recouped the dollar loss by selling 5 April 15 options with a spread order.

At year-end, the price of the stock was $15. We had a paper loss of $2,500, but we had received $6,225 income and spent $3,125 for a net profit of $3,100. If we had not taken a loss in November, our total return would have been higher, but without the tax loss our taxes would have been greater because all gains, with all options, are short term.

Table 8-2: Actions Taken with Falling Prices

500 shares of XYZ Co.
Market price at the beginning of the year: $20.00

Date	Price	Market Action	Income	Cost	Profit/Loss
1/15	20	expired 5 Jan 22½	1,100*		
1/15		sold 5 Apr 20 at 2	1,000		
3/3	18	bought 5 Apr 20 at ½		250	750
3/3		sold 5 Jul 17½ at 1¾	875		
5/24	17	bought 5 Jul 17½ at ¾		375	500
5/24		sold 5 Oct 17½ at 1¾	875		
6/22	16 ½	bought 5 Oct 17½ at ½		250	625
6/22		sold 5 Jan 17½ at 1¾	875		
8/23	15 ½	bought 5 Jan 17½ at ½		250	625
8/23		sold 5 Jan 15 at 3	1,500		
11/29	16	bought 5 Jan 15 at 4		2,000	–500
11/29		sold 5 Apr 15 at 5**	2,000		
	Total		$8,225	$3,125	$2,000

*Last year's option expired this year.
**Next year's settlement.
In addition, there was $300 in annual dividends.

AN ACTUAL PORTFOLIO TRANSACTION

Table 8-3 illustrates the operational ideas we have studied. It shows the actual transactions made in my covered call income portfolio from 8/27/90 to 5/9/91. All dollar amounts have been rounded, including sales commissions.

Remember, margin interest and commissions are paid by the dividends generated from the stocks in the account.

Schering Plough Corp. stock came into my portfolio in 1986 when it bought out Key Pharmaceutical. I had found "Key" on the option page in January 1984 and it was a good write. After further study, I decided it was an excellent underlying stock to hold. In February 1984, with a buy/write order to my broker, I bought 500 shares at $11 and simultaneously wrote covered calls. In September 1984, the stock price dropped to a new low and I reacted. I bought back the November 10 calls and also bought 500 more shares at $9. This lowered my net cost per share from $11 to $10

(500 at $11 and 500 at $9, $5500 + $4500 = $10,000 for 1,000 shares, or $10 per share).

In 1986, the 1,000 shares of Key in the stock swap became 342 shares of Schering Plough. I sold the 42 shares at $36, as I can only use round lots to do options. I always try to keep equal amounts of money invested in the stocks in my covered call income portfolio, around $10,000 per company. I now had 300 shares of Schering Plough at $36 each = $10,800.

Schering Plough prospered through the next few years. It had two two-for-one stock splits. The first gave me a total of 600 shares and the second split a total of 1,200 shares. The dividends increased each year as well.

Now to the period I am using for this illustration. On August 27, 1990 when the market price was $47, I wrote a covered call option for Schering Plough February 45 and received $4 each or $4,800.

The cash call option premium would be taxed as 1991 income, since the covered call expired in February 1991. I had the use of the money in August 1990, though the taxes were not payable until 1992. How's that for a tax deferment!

Table 8-3: The Schering Plough Example

Date	Market Action	Income	Cost	Profit/Loss
8/27	sold 12 Feb 45 at 4	4,800		
1/8	bought 12 Feb 45 at $3/4$		900	3,900
1/8	sold 12 May 45 at $1\,7/8$	2,250		
5/9	bought 12 May 45 at $8\,3/8$		10,050	–7,800
5/9	sold 12 Nov 50 at 6*	7,200		
	Total	$14,250	$10,950	–$3,900

* Settlement after August (one-year period).

On January 8, 1991, the market price of the stock was $40.50. With a spread order I bought back the call at $0.75, a total outlay of $900. This meant I had ($4,800 – $900) $3,900 in capital gains. I then sold 12 May 45 calls for $1.875 each = $2,250.

As you see, I never hold stock without writing covered calls. Yes, the stock could go up and I could write covered calls on it for more money, but it could go down and I would get less. Believe me, it's better to have written the covered call option on a stock and watch it go up than not to have written the covered call option and watch it go down.

The stock did go up, and how! The market price of Schering Plough was $53.50. I knew I would be exercised shortly as it was a May call, but I wanted to keep my low-cost basis in the stock. On May 9, 1991, using a call/spread order, I bought back 12 covered call closing contracts of Schering Plough May 45 for intrinsic value only, at $8.375, or a total of $10,050. That meant a short-term capital loss of ($2,250 − $10,050) $7,800, which went into my capital loss account to be used against other income in my portfolio.

I had a cost basis of only $10,800 in the Schering Plough stock and put in an additional $7,800. You may feel that putting in an additional $7,800 to buy back the expiring May 45 call is a risky business practice. Surely a stock that can rise rapidly from $40.50 to $53.50 from January to May (4 months) can fall just as fast. I judged otherwise.

With the call/spread order I made an opening call sell to provide some money to offset part of the $7,800 May covered call buy back and to provide a partial, risk-reducing hedge against a stock decline from the new high of $53.50. It's unlikely that anyone would want to commit $7,800 for the buy-back without a cash infusion from a new covered call sale. By selling 12 contracts of Schering Plough November 50 at $6, I received $7,200 to go against that loss.

My cash outlay at this point is $600. The debit balance in my margin account has increased by $600. For this $600 debit I have realized a capital loss of $7,800, which can be used in offsetting other income in the account. Any loss remaining can offset up to $3,000 of ordinary earned income. In this example, the covered call option loss is $7,800, a realized short-term capital loss.

As you have seen, it's not an equity loss. I did not lose $7,800. I did not lose anything. I made money in the related transactions, including income tax savings. Taken together, this realized capital loss has a value, at the 25 percent income tax level, of $1,950 in tax savings.

By doing the last transaction I have taken the value of Schering Plough shares from $45 to $50 per share or $1,200 x $5 = $6,000.

The share price of Schering Plough could drop from $50 to $44 before I would begin to lose equity, $50 − $6 = $44. The share price could rise from $50 to $56 before I would begin to lose the upside price movement, $50 + $6 = $56. With the use of covered calls I am protected up or down from $56 to $44. It's a safe, secure, and a profitable feeling to have this protection.

I did not plan the above scenario, nor could I. You have to learn to react to the irrational market system and use it to your advantage.

To make a long story short, on September 4, 1997 I had 4,000 shares of Schering Plough (thanks to stock splits) priced at $50.25 (post split).

On May 9, 1991 I owned 1,200 shares of Schering Plough with a strike price of $50 per share in my covered call income portfolio or $1,200 x $50 = $60,000.

On September 4, 1997 I owned 4,000 shares of Schering Plough with a market price of $50¼ per share or $4,000 x $50.25 = $201,000.

Reducing and controlling risk to the portfolio is the theme of this book—but making money and savings on taxes are also okay!

Give me the luxuries of life and I will willingly do without the necessities. Frank Lloyd Wright

What we call luck is simply pluck, and doing things over and over; Courage and will, perseverance and skill, are the four leaves of luck's clover. The Four Leaf Clover, Unknown

CHAPTER 9
THE COVERED CALL
PORTFOLIO AS A
TAX SHELTER

In 1790, the nation which had fought a revolution against taxation without representation discovered that some of its citizens were not much happier about taxation with representation. Lyndon B. Johnson

What you have left after taxes is your real income. By employing the techniques described in this book. you can reduce, defer, or eliminate taxes on investment income.

Investments that produce only income are exposed not only to inflation but also to full taxation. Protect yourself from the tax consequences of your success. If you have even modest income or profits, you must consider tax planning and tax sheltering. Tax factors will affect your buy and sell decisions in operating a covered call income portfolio. In Chapter 8 we discussed a few strategies you can apply in your portfolio to avoid taxes. This is a more in-depth discussion.

INVESTMENT TAXATION DEFINITIONS

So that there is no confusion, let me give you the accepted definitions of some important terms:

Capital asset: an asset that is not bought or sold in the normal course of business. The IRS considers both stock and call options to be capital assets.

Capital gain: the amount by which the proceeds from the sale of a capital asset are more than the cost of acquiring it.

Capital loss: the amount by which the proceeds from the sale of a capital asset are less than the cost of acquiring it.

Capital loss carry forward: a capital loss that exceeds capital gains and the allowed annual limit of $3,000 against ordinary income. It may be carried forward to subsequent years as an offset to capital gains or ordinary income. There is no limit to the amount of capital losses that may be used to offset capital gains in any one year, but only losses exceeding gains may be used to offset ordinary income.

Cost basis: the original price of a stock, including brokerage commissions.

Earned income: income from wages, salaries, bonuses, and commissions generated by providing goods or services.

Fungible: things of identical quality that are interchangeable. (Commodities such as soybeans or wheat, common shares of the same company, and dollar bills are all familiar examples.) A fungible unit's any unit that can replace another unit to discharge a debt or obligation.

Fungibility: the interchangeability, e.g., of listed covered call options by virtue of common expiration dates and strike prices. Fungibility makes it possible for buyers and sellers to close out their positions by using offsetting transactions through the Options Clearing Corporation (OCC).

Long- and short-term: the holding period required to differentiate short-term gain or loss from long-term gain or loss for tax purposes. Long term currently is 12 months.

Offset (accounting): the amount equaling or counter-balancing another amount on the opposite side of the ledger. Capital gains can be offset by capital losses.

Offset (options): the purchase of an equal number of contracts identical to those previously sold, resulting in no further obligation.

Ordinary income: income from the normal activities of an individual or business, as distinguished from capital gains from the sale of assets.

Realized profit or loss: the profit or loss resulting from the sale or other disposal of an asset. If you sell the asset at a gain, you will have a realized profit. Until you sell, your profit is unrealized.

Tax avoidance: the reduction of a tax liability by legal means. For example, investors who itemize deductions may avoid some taxes by deducting the cost of this book.

Taxable event: any sale that results in a profit or loss that would affect taxes.

Unearned income: individual income, such as dividends, investment interest, covered call premiums, and capital gains realized from invested capital, that is not produced by personal effort.

THE COVERED CALL INCOME PORTFOLIO AND TAXES

Capital gains or losses can result from many activities, such as the sale of stocks, covered calls, real estate and other items. Once a taxable event results in a capital gain or loss, it may be united with all other capital gains and losses for tax purposes. The IRS requires you to net or offset gains and losses against each other to produce a net capital gain or net capital loss for your annual taxes. Long-term and short-term gains and losses must be totaled and then netted out against each other. Your payback may be lower taxes.

If your losses outnumber your gains, your net capital losses can be used to reduce ordinary income to the extent allowed by the IRS. Capital losses can be offset dollar-for-dollar against not only capital gains but also $3,000 of ordinary income.

The 1997 Act reinstated preferential tax treatment of long-term capital gains for certain taxpayers by fixing a maximum tax rate of 20 percent on net capital gains (net long-term capital gains minus net short-term capital losses). Thus, in some but not all cases, individual investors with profitable positions may have an incentive to hold their positions for an extended period.

The long-term holding period is currently one year or more. If stock is acquired and held for more than one year, the resulting gain or loss on a sale is a long-term capital gain or loss. If the stock is purchased and sold in one year or less, any resulting gain or loss is short-term. The U.S. Congress may change these provisions.

The capital loss carryover can be used when you have a greater capital loss than you are allowed to deduct for the tax year. This excess of unused capital loss is carried-over to the next tax year. In this way accumulated capital loss can be used, even if it takes several years.

Avoid unpleasant tax surprises. Keep careful track of both gains and losses so there is still time for year-end transactions to restore any imbalance. If there is a net gain, it's advisable to take any likely loss before year-end to balance against it and reduce or eliminate taxes.

Tax planning requires knowing where you are concerning taxes, and what tax liability will be incurred from your investment transactions. Any investment strategy that ignores tax consequences is not a good one. Tax planning is for all the time for all investors, not just the wealthy; don't wait till April 15th to do it.

COVERED CALL CONTRACT—CLOSING TRANSACTION

When a covered call contract (opening transaction) is sold, neither the profit or loss nor the tax consequences can be determined until the contract ends. There are, as by now you know, three possible outcomes:

1. *Exercise*. The holder of the covered call contract calls your stock away, meaning you have to sell the stock at the strike price to which you had previously agreed.

2. *Expiration*. The holder of the covered call contract did not call your stock because the stock price of the stock did not meet the strike price. The expiration date has passed, and you keep the stock as well as the premium.

3. *Purchasing an offsetting covered call*. Buying a covered call contract to close one previously sold fulfills your obligation to deliver or sell your stock on exercise, allowing you to keep your original stock without incurring a significant capital gain tax liability.

Until the opening transaction (the selling of a covered call) has ended by exercise, expiration, or purchase of a closing offsetting transaction, the covered call will remain open. Premium money received is not considered taxable until the covered call ends. Until that time it cannot be determined whether the covered call contract has produced a capital gain or a loss.

When a covered call is not exercised, the premium is a short-term capital gain. If the covered call is exercised, the premium plus the strike price received become the sale price of the stock. The resulting gain or loss depends on the holding period of the underlying security used to satisfy the assignment. If you have owned the stock for over a year, you will have a long-term capital gain or loss.

Gain or loss on buying a covered call offset closes the covered call obligation as short term, regardless of the length of time the call was outstanding.

If you have stock that has appreciated in value but want to defer the gain until the following year, consider writing a covered call with an expiration date next year. If it expires or the buyer of the covered call doesn't exercise it until next year, both the amount you received from selling the covered call and the proceeds from selling the stock are not reported until the following year.

For example: You receive a covered call cash premium when you sell a covered call in 1998. Because it expires in 1999, it's not reportable to the IRS until April 2000, when you're doing your 1999 tax return. How's that for tax deferment?

TAX GAINS OR LOSSES

Gain or loss in some cases can be determined by the writer of covered calls, namely, you. When you have a covered call gain or loss, the covered call is treated as having been sold or exchanged on the date it ends.

For example: Pete purchased 100 shares of ABC stock for $20 per share on November 22, 1999. On December 1 the stock was selling for $50 a share, but Pete wanted to defer the gain until the following year and protect himself against a market decline. Pete wrote a covered call for $40 per share expiring in three months. He received $11 per share for selling this covered call. Pete has acquired protection against a market decline (he has $11 premium in his account). If the buyer of the covered call does not exercise it, Pete reports the $11 per share as a capital gain. If the covered call is exercised, the $11 per share is added to the $40 per share exercise price, making a total sale price of $51. A gain would have been realized when the covered call position was closed. Total gain is $31 ($51 − $20) per share. Either way there is no taxable event till 2000—and no need to report till April 15, 2001.

If the ABC stock had continued to appreciate, the covered call would have been exercised. Pete could have bought other shares of the stock in the open market to deliver against the covered call, or he could choose to prevent an exercise by purchase or buy back of the call option. A loss would have been realized when the covered call position was closed in either of these two ways. Buying back the offsetting call option contract creates a capital loss, a taxable event. It also produces a gain in equity, which is non-taxable.

Buying new shares to deliver to a covered call assignment forces a choice. The buyer of the call option doesn't care how long the stock was held, be it 10 years or one day. All he wants is the stock you're obligated to deliver to him at the strike price. Since all common stocks are fungible, you can deliver old or newly acquired shares.

This is where you can choose whether you want a gain or a loss on providing these shares of stock.

If the cost basis is higher than the current market price, buy new shares at the lower price and deliver the old, higher-priced, shares. Now you will have a lower cost basis for the shares in your portfolio and at the same time have a larger realized loss on this transaction.

If the cost basis is lower than the current market price, buy new shares at the higher price and deliver these, keeping your lower-cost shares in your portfolio to avoid a capital gain. Delivering the new, higher-priced, shares will give you a larger realized loss on this transaction.

Of course if you're carrying over a capital loss offset, the lower basis stock could be used for delivery in both cases and a larger capital gain would be realized for these transactions.

Keep a running total on the capital gains or losses in your covered call income portfolio. Only if you have this information available can you determine which of the closing covered call strategies to use. If there are losses, take gains; if there are gains, take losses. Remember, capital gains plus up to $3,000 of ordinary income can be offset by capital losses, dollar-for-dollar.

For all profits and losses realized as short term you will pay the highest tax rate.

Think in terms of your total tax liability. With careful planning, you can realize a long-term capital gain on one side and a short-term loss on the other.

A tax shelter program helps to reduce, defer, or eliminate taxes on personal income. While the program may offer reasonable economic gains, the first and second possible outcomes of a covered call are both taxable events, because we retain the premiums after the covered call option contract expires. The third possible outcome is not a taxable event. Unrealized gains on the increase in the value of a stock are non-taxable.

There can be tax benefits when the covered call is exercised. The length of time you held the underlying stock decides the holding period, not the covered call option. When you own the stock short term, the tax consequences of exercise are short term. When you own the stock long term, the tax consequences are long term. The premium is considered part of the selling price of the stock. It adds more net gain to the transaction (and lowers the real, out-of-pocket cost).

For example: When ABC Corp. is bought in January at $20 (cost basis), the investor writes an April 20 covered call for a premium of 2, making the out-of-pocket cost $18. If the call is exercised in April for $20, the investor has a $2 short-term gain. Thus, with call options the premium increases the amount realized by the writer on the sale of the underlying stock.

Now let's assume that the investor has owned ABC Corp. for several years. In January the investor writes an April 20 covered call for a cash premium of 2. If the stock is exercised in April, the tax sale price is $22 ($20 cost + $2 premium). The $2 premium profit as well as any amount above the original cost of the stock is a long-term gain.

At year-end if the price of the stock has increased the investor buys back any pending covered call to take a loss on this year's tax return. He then sells a new covered call with an exercise date in the next year. This premium will not be taxable until the next year's tax return, after the covered call has ended. By buying back the covered call, the investor could also extend the holding period of a short-term underlying stock until it becomes a long-term holding.

Taxable events can be decreased and non-taxable events increased with a high degree of control in a covered call income portfolio. Calls are an excellent means of tax-sheltering income.

Most writers in the financial press never discuss the opportunity to use exchange-traded covered call options. The buy-back capability exists because of the OCC. In the prospectus of the OCC you can study the mechanics. The OCC makes the covered call contract fungible. All call options for the same underlying stock having the same exercise price and the same expiration date are interchangeable.

If you buy back fungible call option contracts regularly, it will benefit both your covered call income portfolio and your tax position.

Patience is not only virtue, it pays. B.C. Forbes

The one thing that hurts more than paying an income tax is not having income to pay income tax on. Harvey C. Friedentag

CHAPTER 10
CONCLUSION

One can survive everything nowadays except death. Oscar Wilde

Your reason for investing is to make money. Successful investing is like marshaling forces on a battlefield; all the pieces must be working together. Tactics that are not aligned can produce chaos over a long period of time. A well-thought-out plan and goals, generally executed with discipline, can be undermined by investing in the wrong stock at the wrong time, or by dollars lost through taxes and inflation.

Your *investing* must be business-like. In spite of running a successful business operation, some capable business people operate their investment portfolios with complete disregard of sound business principles and practices. Making profits from investing is a business venture and requires the development of sound business habits.

Know what you're doing. Know as much about investing as you know about the business in which you made your original capital.

Operate your investment business yourself. Have the conviction that you know what you need to know. If you have formed a decision based on the facts and you know that your judgment is sound, act on it. If you make a decision but do not act promptly on it, you have lost your time and opportunity. Act timely!

You are not necessarily wrong if the crowd disagrees with you. You are right because your research and reasoning are right. It is not what you think about a stock, it is what you *know* about a stock.

The main thing to remember with the stock market is never to put your money in just one company. Do your homework before you invest. Choose carefully. Spread your risk over a portfolio of stocks. Become an authority on covered call writing before you sell any calls. There is no equal for knowledge. It is both power and security.

To be sure, we cannot possibly know all the ins and outs of investment markets. Most of us made our money making products or selling
services. Making money with money is an art that has to be learned—
often the hard way.

FEAR, GREED, HOPE, AND ONE'S SELF

What prevents us from investing systematically, applying all of the
lessons learned from our past mistakes?

The answer stems from the same elusive reason that lead us to break
New Year's resolutions, have extra helpings when on a diet, and never get
around to fixing that leaky faucet—human nature. The greatest obstacle
to successful investing is lack of discipline and failure to act.

Two enemies of the average investor are fear and hope. These are
usually accompanied by another—greed. Failing to control emotions
causes us to keep a losing position too long, take profits too early, or take
advice from uninformed people. Controlling these hindrances to success,
which exist in varying degrees within all of us is essential for successful
investing.

With control of emotion and investment knowledge you will not
merely hold your position but will make major capital gains. It should be
repeated that investments are good in any economic climate.

You need not fear during a panic or market crash. You can make an
ally of panics and crashes by understanding their nature. You can avoid
the mental upset and the emotional trauma—if you grasp the patterns and
learn the clues. You can preserve your capital while those around you are
selling in fear when they should be buying, or buy in confidence in a
declining market.

Follow these simple guidelines and you'll develop the traits of a
winner.

- Pay attention to the market. Scrutinize daily market prices, news,
  and trends.
- Have enough capital to absorb reasonable losses.
- Maintain long-term purchasing power in your margin account.
- React to the market and the buy/sell signals with no hesitation or
  emotion.
- Diversify. Trade in 15 to 25 varied industries; never hold more
  than two companies in the same industry.

- Establish a well-tested, long-term trading system.
- Keep up with the latest information. Have reliable, current sources for research.
- Avoid margin calls.
- Plan a strategy. Develop sound business practices. Cut losses and let profits run. Always look for the bargain, undervalued stock.
- Control risk. Use covered call selling for protective hedges on all equity positions in the account.
- React promptly to what you learn about the market.

These characteristics are important in achieving success in the option markets. Nothing prevents an amateur from taking on the traits of the winners. Remember that

1. When stocks are moving from under- to overvaluation, total returns are so high that a fully invested strategy will almost certainly produce superior returns compared with a market-timing approach.
2. Contrary to popular belief, stocks can be good investments even when they are expensive, as long as the overvaluation trend is intact. But when they are overvalued and the correction has begun, those stocks must be sold.
3. Money and inflation trends are critical. The ideal breeding ground for a mania is when inflation is falling and money growth is accelerating. Liquidity, rather than business activity and general prices, pushes stock prices up.

Surprisingly, even during bull market manias, most investors lose money. Why? Because they cannot cope with the high level of volatility and risk that is an inherent element of the mania climate. However, investors who understand the mania process and who can follow a sound strategy can capitalize on a unique opportunity. They can profit handsomely while managing and minimizing the risks by practicing the strategies I have discussed.

Note: A long-term, non-speculative, disciplined strategy is essential. To endure the high volatility possible in any environment requires patience and determination.

We have looked at the basic principles of investing and the advanced usage of these ideas. If you expect to succeed you must have a goal, a plan, and discipline. Lack of any one of these will seriously hamper your success in any investments.

The only way to tell if you're getting anywhere in the long run is to see how you're progressing on the road toward your target. So, if you're really serious about this business, the first thing you have to do is establish goals and objectives. Then plan how to reach them.

I conclude this book by saying the best of luck to you in your stock market adventure. You earned your money the hard way and you now have a safe, methodical way of investing—and a solid foundation for building your wealth. You now have the knowledge to strip the mystique from investing. You now have a clear method from the very basics to a degree of sophistication on a par with the professionals. On the way, you learned to avoid the traps that constantly frustrate others who attempt to make a fortune in the stock market.

Don't gamble! Take all savings and buy some good stock and hold it till it goes up, and then sell it. If it don't go up, don't buy it. Will Rogers

The one who shoots best may sometimes miss the mark; but the one that shoots not at all can never hit it. Owen Felltham

APPENDIX I
SOURCES OF
INFORMATION

I bought stocks like they were going out of style. And they were.
Anonymous

FINANCIAL PUBLICATIONS

Choosing from the army of financial periodicals assaulting your mailbox and wallet these days can be tougher than picking a stock. And the list just keeps growing. For the average investor who just wants general information, *Money Magazine* and *Kiplinger's Personal Finance Magazine* are the sources to read. If you're a little more sophisticated, the information found in *Forbes*, *Business Week*, *Financial World* and *Fortune* magazines offers more.

Two publications, the *Wall Street Journal* and *Investor's Business Daily*, keep readers abreast of what's going on in the world and in business. They both print stock tables, but these publications also give investors a lot more information.

Read the business stories on the financial pages. The information reported, such as dividends, earnings reports, new products and services, takeovers, and selling of assets, is extremely valuable. The business press publishes abbreviated press releases before companies mail annual or quarterly reports to investors. A newspaper story contains the meat of a company's quarterly performance. Frequently they add the insight of Wall Street analysts who follow a company to decide its potential investment value.

Worth magazine is a well-written and illustrated bimonthly publication that covers investing topics in a varied and delightful way. *Better Investing Magazine* is the official publication of the National Association of Investors Corp. (NAIC). It covers fundamentals in a clear and concise way, teaching as it goes.

When information about a stock in the daily press catches my attention or as a customer I like a company and its products or services, the first thing I do is open my *Standard & Poor's Stock Guide*. The *Guide*, published monthly, is an excellent starting point. It covers more than 9,000 companies.

Perhaps the only thing you know about a stock is its name. By using the *Guide* you'll find out if the company is publicly traded (it may be privately owned), if it's optionable, its stock symbol, its main line of business, and its Standard & Poor's rating. Also included will be the company's price history for the past decade, the previous year, the current year to date, and for the past month.

Value Line Investment Survey, published weekly, covers 1,700 stocks. In its Summary and Index the *Survey* presents up-to-date rankings for price performance, timeliness, and investment safety. The stocks are rated from 1 (highest) down to 5 (lowest). In addition, each of the 1,700 stocks is the subject of a comprehensive, full-page Rating and Report at least every three months. It also carries analyses of interesting industries to look at.

Barron's, published weekly, is another exceptional source of investment data and always lists the last reported earnings of a company. In the *Wall Street Journal* and *Investor's Business Daily*, earnings are announced by a company and reported only once.

Many investors choose to use *Standard & Poor's* and *Value Line* as their primary sources for evaluation. They use other sources for additional background information. Before you spend time studying any single stock, use *Value Line* and read the general discussion of an industry to get a good overview. Then read the discussion of particular stocks in that industry. You may find another stock in the industry that is better than one you had in mind.

Forbes and *Financial World* are two of my favorite magazines. Published bimonthly, they contain an enormous amount of ideas and information on companies and the economic climate. The editorials and investigative reporting in *Forbes* are wonderful reading. *Forbes* annually rates companies within industries, comparing their profitability and performance. *Financial World* offers an independent appraisal of 3,000 stocks bimonthly. Both magazines have mail-in cards to request annual reports from companies in which you may be interested.

TRADE PUBLICATIONS

One of the very best sources of information is known as the trade press. For anyone concerned with the wide world of investments, these narrowband periodicals can often be worth their weight in gold. The intense focus on an industry or slice of business provides information that rarely makes the dailies or newsweeklies. Even when it does, it's long after such information appears in the trade press. For instance, if you had read *American Banker* during the late 80s and early 90s, you would have known in advance of the disasters that have struck the banking industry. If you're looking into as the medical field, it's a good idea to pick up *Medical Economics* and page through it.

Since the computer and all its peripherals have become actively represented in the stock lists, I've made electronic news part of my perusal routine. *Occupational Hazards: the Magazine of Safety, Health and Environmental Management* keeps me posted on important matters in this industry. *AV Video Production and Presentation Technology* writes on matters of interest to their segment of the communications industry. Through it, I find companies that are the leaders in this fast-developing field.

The trade press is also one of the most important sources for finding the "up and comers." Reading the editorials and industry concerns will give you inside information you'll need. The advertisements will give you the names of companies to look at. I send for their annual reports and study them. I ask people in those fields questions about their products. Ask your doctor, lawyer, tradesman, and friends which products they like and use in their own occupations. The opportunities can be endless.

Many of these sources should be available at your public library. Many people are afraid to ask for help, but the business or reference librarian is there for just that reason.

INVESTMENT CLUBS

Joining an investment club can be a safe and reliable way of investing in stocks and learning by doing. There is a non-profit organization called NAIC (the National Association of Investors Corp. which has 9,200 investment clubs nationwide. The NAIC can help you find a club in your area. Call or write the NAIC at (313) 543-0612, 711 W. Thirteen Mile Rd., Madison Heights, MI 48071.

The NAIC offers investment clubs an investor's manual and seminars to provide guidance in the selection of securities and the management of an investment portfolio. Investment clubs do permit people to put small amounts of money into the stock market, in some clubs as little as $10 per month. The average club has about 12 members and a portfolio of about $100,000 invested, or $8,000 a person.

Joining an investment club is a way to learn, and a way to earn. The clubs will expect some sweat equity from you. Each member may be responsible for finding information about a company and recommending whether to invest in it or not. Meetings are held once a month. Once you join a club, it should be a long-time commitment. The best part of the meeting is discussing the club's portfolio and whether to buy, sell, or hold different investments.

Investment clubs work. Mark Hulbert writing in *Forbes* (November 25, 1991) stated that "groups of individuals have done so much better than the professional investors." Many clubs, using a long-term approach and investment savvy, have beaten the market.

Because we live in a world of daily financial decisions, investment education is a lifelong process. Many individuals get their first successful stock investment experience in an investment club. I belong to such a club because the monthly meetings with a group of individuals interested in stocks builds my investment exposure. The club helps me to follow the latest stock market trends and opens the door to likely investment opportunities. More minds and people of different backgrounds and interests expand the range of investment ideas.

The low-cost investment plan offered by NAIC is worth the price of the nominal annual dues. It teaches how to buy stocks with low or no commissions, how to participate in dividend reinvestment plans, and where to find general education on undervalued stocks that will outperform the market. There are over 150 companies whose stocks can be purchased directly through the NAIC, avoiding the beginner's hurdle—high commission costs for buying a few shares.

Forget the past; no company has yet backed into success. Anonymous

GOING ONLINE

Currently, if you want investment information, you need search no further than your computer screens, where you can bring up the vast resources of the World Wide Web. From stock and market data to financial and on-line brokerage accounts, there is something worthwhile for you somewhere on the Web.

The Internet is a choice investigative tool, perfect at putting facts at the fingertips of investors who need it. The Internet will do whatever you want in the way of information searching, and it does it fast and efficiently, giving you the right information.

However, finding the information is another matter. While many sites provide in-depth financial statistics, data, and other information, other sites are purely promotional, with little or no useful value. Some sites offer free access to information; others charge a fee.

About a third of all U.S. households now have personal computers with modems. It's estimated that the number of people with online trading accounts will approach 1.3 million by the end of the year and reach 10 million in 2001. The modem is a shot across the bow of every traditional stockbroker, financial planner, and mutual fund.

Investors don't need intermediaries to get them investment information or take their orders. You don't have to make phone calls and wait on hold for "the next available representative." You can do it all on the Internet, 24 hours a day. Investing by Web is not only convenient, it's cheaper for those who buy stocks and other financial products that carry sales commissions.

A knowledgeable online investor can visit *www.freeedgar.com* to retrieve real-time Securities and Exchange Commission filings for companies, then hyperlink straight to the financials, to management's discussion of the competition, or to full explanations of those unfortunate lawsuits. You can search the Silicon Investor message board, *www.techstocks.com*, for the call on technology stocks. You can use Market Guide, *www.marketguide.com*, to obtain the latest earnings reports and historical financial data for companies.

With all that instantaneous third-party information, why bother with corporate Web sites?

You should and here's why: Internet use doubles every few months; 16 percent of American adults have access to an online service or the Internet. About half go online every day, and many count investing among their top priorities. When they begin surfing for investment information, chances are most won't find their way to Freeedgar, Silicon Investor, or Market Guide. Most will go to Yahoo! or Excite, type in the name of the company in which they're interested, and go straight to the corporate site.

When I surf the World Wide Web, my interest is far from casual. One favorite stop is the EDGAR database of corporate filings, hosted by the Securities & Exchange Commission. I keep an eye out for SEC form S-1s, which often disclose insider selling. This is the kind of sophisticated security analysis that pros have been doing for years, using expensive databases and SEC-document retrieval services. I don't use the costly services for my investigative work; I use the Internet.

Today, online resources are huge, and increasingly user friendly. The message boards of the Internet have become "must reads" for investors.

Web sites sponsored by companies, publications, and exchanges are among the multitude of research tools that can provide invaluable resources for investors. Companies welcome the Web as a way to free themselves from the tyranny of institutional buyers and to appeal to the individual investor, who in theory will be a longer-term investor.

Most company Web pages include a current annual report, information on direct-purchase and dividend-reinvestment programs, and a listing of telephone and mail contacts for further information. A handful of company sites deserve acclaim for their quick-loading graphics; compact lists of links that save users from having to scroll down screens; long documents broken down into easy-to-swallow topic areas, with hyperlinks to the most interesting information; quotes as well as price and volume charts; one-click access to all recent SEC filings; and an invitation to send comments and inquiries via e-mail. The best sites also provide something extra, data or information not readily found on third party sites, providing one more good reason to invest in the company.

Web-wise, I gather useful information through the public company directory at *www.investorguide.com*. Type in the stock symbol in the search field, and the clever directory will assemble links to the best in current quotes and charts, third-party profiles and research, and all pertinent SEC documents. I think of it as my all-purpose, do-it-yourself investor-relations site.

Useful Web Sites

To visit the sites I have listed, you need a computer with a modem, an Internet service provider, and a Web browser.

Once connected to the Internet, type the Web URL for the site you're looking for into the address window at your service provider's home page and press enter. This takes you to the home page of the site. The baud rate (speed) at which your modem connected to the Internet will determine how fast the home page is brought up onto your computer screen.

You can explore the contents of the site. If you're not sure about who's responsible for the information presented at a particular site, check the bottom of the home page for a link to a company or person, or look for an "about" icon. Not all sites will be clear as to where the information is coming from, so be wary. Do not accept information as fact just because it has appeared on your computer screen.

When exploring a site, it's easy to get lost after clicking on several icons or links that take you to different pages on the site or take you to other sites. Web browsers provide "back" and "forward" buttons so that you can retrace your steps or return to a particular page. You can tell if a click on your mouse has taken you to a link outside of the site by checking the URL in the address window; it will be either a long extension of the original URL you entered or a completely different URL.

I have sorted through various sites to weed out the junk, and determined their value to save you an enormous investment of your time. The alphabetical list which follows has undergone my scrutiny and includes the free sites I personally use.

Keep in mind that Web sites are prone to frequent changes and the details I provide may not match what you find by the time you reach a site listed here.

AMERICAN CENTURY   www.americancentury.com
Stock, funds, options with quotes.

BUSINESS WEEK   www.businessweek.com
Business Week, finance news, quotes, and analysis.

DATA BROADCASTING CORPORATION   www.dbc.com
Stock and fund data, quotes and news.

FINAL BELL   www.sandbox.net/finalbell/pub-doc/home.html
Does a bull market make monkeys of brokers?

HOOVER'S ON-LINE   www.hoovers.com
Company profiles for stock research.

INVESTOR'S BUSINESS DAILY WEB EDITION   www.investors.com
Financial news and analysis.

INVESTORS EDGE   www.irnet.com
Provides current information normally used by insiders.

MARKET GUIDE   www.marketguide.com
Stock data, news, and quotes.

MORNINGSTAR NET   www.morningstar.net
Stock and fund data, quotes and news.

NASD REGULATION   www.nasdr.com
How is your broker behaving? To be sure, go here.

NETworth   www.brill.galt.com
A great free site to get quotes and company information.

PR NEWSWIRE   www.prnewswire.com
Finance news and quotes.

RISKVIEW   www.riskview.com
Investment-risk analysis. Here you can find out how risky your investments are.

S&P EQUITY INVESTORS SERVICE   www.stockinfo.standardpoor.com
News headlines and stock picks from Standard & Poor's.

STOCK RESEARCH GROUP   www.stockgroup.com/index.html
Up-to-date stock and investment information.

STOCK SMART   www.stocksmart.com
Portfolio tracking and quotes.

STOCK WIZ   www.stockwiz.com
Stock data, news, and quotes.

USA TODAY MONEY   www.usatoday.com/money/mfront.htm
A comprehensive assortment of news and data.

WALL STREET CITY   www.wallstreetcity.com
Stock screening, news, quotes, and more.

WALL STREET JOURNAL INTERACTIVE EDITION   www.wsj.com
Finance news, quotes, and analysis.

YOUNG INVESTOR   www.younginvestor.com
The entertaining fundamentals of money and investing.

If you pan through the sands of the Internet, you can really find valuable nuggets of investment information.

Every day, from CNBC to the Internet, a tidal wave of financial information hits investors. Don't get swept away. It won't be easy. Investors today are culling investment tips from a dizzying array of sources. A recent survey showed that 69 percent of investors obtain financial news from daily newspapers; 63 percent read business publications; 38 percent watch financial television shows, with radio listeners following closely at 37 percent, and 45 percent surf the Internet for stock-related news.

Unfortunately, much of what investors are finding is useless. Separating hype from valuable financial information is trickier than ever. The key to filtering the news is to match your investment goals with the kind of information you're receiving. If you're holding a stock for the long haul, for instance, don't get unhinged by bad news that has no long-term strategic implications.

In the short run, the market is a voting machine. But in the long run, it's a weighing machine. Warren Buffet

COMPANY REPORTS

The Annual Report

Company annual reports are progress reports to the owners of a company—the shareholders. Companies listed on the New York Stock Exchange must submit an annual report to each shareholder. The report provides valuable information on the state of the business in the past year, and the outlook for the future. It's better than the grapevine, news releases, or rumors.

Your research begins with the financial statements contained in the annual report. Annual reports are an investor's window to a company. They are as revealing for what they don't say as what they do. Unfortunately, lacking a financial background, most investors tend to concentrate on the letter from the chairman, browse through the feature articles, and look at the photographs.

The annual report contains a lot of data and is the place to begin looking in order to study a company. It doesn't take much effort and you can save yourself possible future problems and losses.

Annual reports are filed 90 days after a company's reporting year ends. They often include a letter from the chairman of the board to shareholders, balance sheets, income statements, and discussions of the yearly performance results.

Certain pieces of information in the report are more important than others. Cut through the promotional material and get to the essentials.

Start at the end of the report. This is where all the numbers and fine print are. *Remember, the large type giveth and the small type taketh away.* In the small type you'll find the report from an independent auditing firm whose task is to look over everything. They make sure that the annual report is honest, thorough, and accurate in recording business and financial transactions.

Look at the Accountant's Opinion Statement. Since financial reporting involves considerable discretion, the accountant's opinion is an important assurance to the investor. If there is a qualified opinion, it signals concerns and warrants further investigation. Throw away the annual report that has a qualified opinion, and look elsewhere for a potential investment.

The financial report must be based on the guidelines published by the Financial Accounting Standards Board (FASB). Where the guidelines are not clear, accountants must rely on their professional body of knowledge and the practices prevailing at the time the decision is made. These interpretations, guidelines, and judgments are customarily termed "generally accepted accounting principles." Be sure the auditor's report gives the company a clean opinion with no reservations.

You can also get a more detailed picture of how a company is doing from its filings with the Securities and Exchange Commission. Particularly revealing is Form 10-K. This is an expanded annual report that details company history, property, competition, and industry outlook. These are available upon request from the company's Investor Relations Department.

The Financials. The Income Statement is a vital piece of information found in an annual report. It shows how much the company made or lost in the past year by adding up sources of income and deducting expenses. The earnings result is the net income, which is often translated into earnings per share. Earnings per share are what savvy investors look for. This shows how well a company's core business did. Beware of one-time gains, such as when a company sells some real estate. The windfall may boost the earnings but will not tell you how well the company is being run.

Check what portion of the corporation's earnings per share goes toward paying cash dividends. Growing companies probably will pay out little or no dividends because earnings are frequently being reinvested in the company. A utility will commonly allocate more than half its earnings toward dividends. Compare this year with last year. If a company is paying out an increasing portion of earnings to sustain dividends, this could be a sign that the business is having some financial difficulty.

Check the footnotes for possible changes in accounting policies during the year. A change in, e.g., inventory or depreciation can have the effect of raising flat or declining earnings.

The footnotes also contain information about pending or possible litigation. Financial time bombs can be hidden here.

Check to see if earnings increased because of higher operating profits, or if the company sold some assets, such as a plant, securities, or even one of its businesses. A comparison should be made of both operating and pre-tax income with sales for several years to detect variances.

Extraordinary refers to an event or transaction during the year that is considerably different from normal activities. What is considered extraordinary? It can be a substantial gain or loss on the sale of a business or asset. A gain could come from the recovery of property or money through a court settlement. You'll find extraordinary items explained in detail in the Notes to Consolidated Financial Statements.

Check the company's Statement of Accounting Policies to find out if the corporation is capitalizing money spent for research and development, engineering, or training. By deferring those charges over several years, the earnings can be made to look better in a particular year, but those items represent deferred costs and are not assets.

Accounts Receivable are the unpaid bills of customers to whom finished goods or services have been supplied. It's normal for businesses to extend credit to customers and receive credit from suppliers, but be sure that there's an allowance for bad debts, since there's always a possibility that customers may not pay their bills. The allowance is based on total receivables and is generally a percentage applied by the accountants. In a difficult economic climate, the percentage is usually higher than when the economy is good.

Other Current Accounts usually include prepaid expenses such as insurance, rent, etc.

Check to see if the ratio of Accounts Receivable to sales is growing. If so, revenues are not coming in as they should. This can suggest poor credit or collection practices.

Inventories are a mixture of raw materials, goods in the manufacturing process, and finished products ready to ship to the customers. Check to see if inventories are increasing. This can suggest either poor purchasing practices or a lack of sales aggressiveness.

The Balance Sheet is an accounting statement showing the company's financial condition. It's broken into two parts: assets and liabilities.

Assets are the resources of a company. They are employed by the management of a company to produce products, services, and sales. Assets usually get divided into current assets and non-current or other assets. Current assets get classified as such because they can be turned into cash usually within one year.

Liabilities are debts owed to others. Current liabilities are debts due within one year. Short-term debt is the portion of debt that must be paid within the year. Accounts payable are the amounts due to suppliers who have extended credit to the company.

Liabilities can be short- and long-term financial obligations. Subtract total liabilities from total assets to get shareholder equity, the portion of the company owned by investors. Shareholder equity typically grows at well-run companies because profits remaining after dividends are reinvested into the company.

Compare long-term debt to shareholder equity. This is a measure of leverage, the use of borrowed money to enhance the return on shareholder equity. The stockholders should have more invested in the company than the lenders do.

The balance sheet summarizes a company's assets, liabilities, and shareholder equity.

There are some basic questions you'll want to ask as you look at the balance sheet. What is the company worth if the business gets liquidated or sold? How does current debt compare to assets? Have there been any large changes in the items listed?

Cash needs no explanation.

Property, Plant, and Equipment refers to the depreciated value of fixed assets. This is the investment management uses to produce, distribute, and sell its products. The amount shown represents the net plant investment after deducting the depreciation allowed by the IRS.

The items included in the Book Value of a company may vary significantly from market value. Equipment or supplies on-hand for making a discontinued product may be worth less than indicated in the books. Yet, buildings and plants that are depreciated on the books can be worth a great deal more than the book figures.

Patents and Other Intangibles include the patents acquired for certain products, trademarks, or goodwill. Goodwill could be the excess over book value paid for an acquisition of another company. (Did they pay too much? Does it fit in with their present businesses?) These assets have no physical existence yet still have a value to the company.

Accrued Payrolls and Accrued Pensions are part of the money the company owes—salaries and wages—to its employees. In addition, these employees have earned credit toward a pension to which the company contributes. Such debts get shown under these headings.

Income Taxes Payable usually get stated separately; these are taxes that are owed but that do not have to be paid until later.

Long Term Debt and Deferred Income Taxes are debts due beyond one year.

A fast check will catch the assets and liabilities. The difference is Shareholders Equity, how much of the company the stockholders own. Total shareholders equity is the corporation's net worth after subtracting all liabilities from total assets.

Par Value is an arbitrary figure and has no relationship to stock value.

Paid-In-Capital is capital received from investors for stock in the company, as distinguished from capital generated by earnings.

Additional Paid-In-Capital is the amount shareholders have paid in over the par value of each share. It may be donated stock, capital, or property.

Retained Earnings can be called surplus earnings. They are the profit left in the business after the payment of dividends.

Footnotes are an integral part of financial statements. They *have* to be read to understand the entire financial report because they are where items that must be reported can be hidden.

Changes in Financial Condition show how much money a company has to work with and identifies sources of cash and disbursements.

The Opening Page. Now that you have done all the important studying, you can continue browsing toward the front of the annual report. Go through all the fancy color graphs, photographs, and charts on expensive paper to the opening page, find the Chief Executive Officer's color picture and ask, "Why is this man smiling?" You hope you can answer the question, "Because he has done an excellent job."

The Cover. When I teach investing, one of my most popular sessions is to scatter a pile of annual reports, and ask the students to look at the covers and decide if the company had a good year or a bad year. Those companies with quantitative items on the front usually had a solid year. Those with modern art and non-business related subjects on the cover tend to have had a poor year.

One problem lies with the idea that there are earnings curves that can rise in unbroken lines. Investors put a premium on such curves, ignoring companies whose curves run flat or follow a fluctuating course. Those without the right curve get hit where it hurts, right in the P/E ratios. A cyclical company, no matter how good it's, is lucky to get 10 times earnings in the market. Those with sweeping up-curves go for 20, 30, or even

50 times earnings. What this does to stock prices, options, merger possibilities, even to corporate prestige is all too apparent.

Imaginative companies may start doing things with their books so that their earnings appear to have those attractive curves, even if they really don't.

Today, most annual reports must be studied skeptically. Many are outright deceptive. Only a minority are truly honest. This occurs despite Opinions handed down by the American Institute of Certified Public Accountants, despite rulings by the SEC, and despite so-called full disclosure by management.

The following two paragraphs cover only some of the cases where confusing and sometimes misleading reporting gets freely used. Once there was faith in these two traditional paragraphs.

"We have examined the consolidated balance sheet of The ABC Co. and subsidiaries as of December 31, and the related statements of income, retained earnings and capital surplus, and the statement of source and application of funds for the year just ended. Our examination was made according to generally accepted auditing procedures that are considered necessary in the circumstances."

"In our opinion, the accompanying consolidated financial statements present fairly the financial position of The ABC Co. and subsidiaries at December 31, and the results of their operations for the year then ended, in conformity with generally accepted accounting principles which have been applied on a basis consistent with that of the preceding year."

To the ordinary investor, and to many professionals, these two paragraphs meant that auditors have checked the figures and found them sound. Then came the scandals. The public learned that the two paragraphs gave no guarantee of the facts upon which the figures were based.

Now the accountants are more careful. For those investors willing to make the effort, annual reports have begun to have more of a between-the-lines meaning.

Now the phrase "subject to" is fairly common in an audited statement. The subject is usually specified in a footnote. It means that the auditor is in serious disagreement with the company about its disclosure or lack of disclosure concerning an important area.

A "subject to" can suggest that assets and earnings are as indicated only if inventories or other assets like investments are really worth what the company claims. This claim the auditors do not certify. What this bluntly means is, "Here are the figures, but we won't swear by them."

For example, consider the following:

"In our opinion, subject to the realization of work-in-progress inventories and accounts receivable described in Note 3, the statements mentioned above present fairly, except for the change in accounting for administrative and general expenses and independent research and development costs described in Note 2."

"The change in accounting in Note 1 increased earnings by $44 million out of a total of $88 million reported."

Don't think you can skip the footnotes just because the accountant has made no objections. Sometimes the only way to understand a company's annual report is to compare it with another report in the same line of business. For example, Ford Motor Company's annual report makes more sense when read with that of General Motors.

This does not suggest that every company is "curving" its books. There are still many companies, du Pont, Eastman Kodak, IBM, Corning Glass among them, that keep conservative books.

You have to learn the meaning of "the quality of earnings." Two companies each report earnings at $4 per share. This does not mean that both are equally profitable. On a strictly comparable basis, one of them might only be earning $3.50 per share and the other closer to $4.50, depending on their accounting procedures.

High-quality earnings are more stable earnings. By borrowing, in effect, from the future, the less conservative company runs the risk of a severe downturn when business weakens. The more conservative company, by contrast, has built-in stability.

Though numbers are still the investment community's best and brightest yardstick of performance, numbers do not always paint the entire picture. Using artwork and visuals, they can be stretched, shrunk, and shaped to fit into almost any mold.

The Chairman's Letter will give you an impression of the company. What happened and why? What is the chairman's insight on economic and political trends? Make certain that the chairman's letter is "up front" in more ways than one. While you're there, check out the financial review that describes how the operating groups and their divisions fared over the most recent three years.

For most serious investors, the main purpose of reading financial reports is to glean some basic historical knowledge about a company, not to discover the next Wal-Mart. Look at the company's annual reports over the last three years to get a feel for how realistic their forecasts have been. Though certainly a corporation's past performance is no guarantee of future results, a company whose forecasts have been reached is likely to have good management. Well-grounded forecasts could make the company's stock a big winner.

The Charts, Graphs, and Photographs. I caution everyone to verify any visuals by returning to the numbers in the back of the annual report, because while numbers don't lie, pretty visuals are not above bending the truth.

Despite how innocent they appear, charts, especially those advanced as a portrait of a company's financial stability, require the closest inspection.

Usually, financial figures are provided by people who have something to gain by how they get interpreted. While companies are legally required to furnish accurate figures, just how accurately they portray the company in visuals is often a matter of perspective.

Change the scale of a chart or a graph. Use a wide-angle lens for a photograph. What is really a financial free-fall suddenly looks like nothing more than a harmless, subtle dip.

Maybe investors, not to mention business people who bought companies based on exaggerated figures, should have paid more attention to the footnotes and the accounting changes in the first place.

The Quarterly Report

I use quarterly earnings as a thermometer to keep tabs on corporate health. Daily headlines about companies' quarterly earnings being up or down from the same quarter a year ago usually have an immediate impact on their stock prices.

Three-month profits can be way up or down for any number of reasons: currency fluctuations, one-time windfall or cost-cutting write-offs, high start-up costs for new products or plants. Some special reasons that could make a quarterly report's results go down, if properly understood, should send a stock price upwards. The bottom-line results are not the sole criterion for judging corporate value and corporate prospects. It's the figuring behind the figure that gives it true meaning.

It's up to us to try to dig out why a quarterly earnings report isn't the whole story. Many managers under quarterly financial press reporting, take to juggling, managing, overemphasizing, and window-dressing their figures. They need to look good, often at the cost of long-range planning for capital investment, research, and development. Quarterly reports are unaudited.

ANNUAL MEETINGS

Even more interesting, if the opportunity presents itself, is to attend an annual meeting. This is the annual gathering of a corporation's directors, officers, and shareholders. At this time new directors get elected. shareholder resolutions get passed or defeated, and operating and financial results of the past fiscal year get discussed.

Most of these meetings are very matter of fact, conducted in a very business-like atmosphere. Occasionally the CEO gets put under pressure while honestly answering shareholder questions and explaining the company's business direction.

A memorable annual meeting for me was when an investor encouraged the CEO to get rid of the "RE" words. Every annual meeting the shareholder attended recently included talk of RE-evaluating the company, RE-structuring it, RE-vitalizing the work-force, RE-assessing the marketplace, and RE-directing resources. Next year, the shareholder suggested that the RE word be RE-ality.

Of course, if the company is RE-ally doing that poorly, the investor should RE-tire from that company by RE-deeming the shares held in that corporation, and RE-investing the money elsewhere.

INSIDER INFORMATION

One guaranteed system universally successful is inside information. For instance, if you know that ABC Corporation is going to show a large and unexpected earnings increase, or that it will be a buy-out candidate at a higher price than will be made public three days from now, buy shares of ABC today; you could be confident of seeing them rise after the announcement.

Inside information is the very best of stock advice. But there are two problems with it. First, it's illegal to use this information to make trades.

Second, inside information is just that, inside, and is usually not available to an outsider.

By the time an individual investor gets to hear about it, it's no longer inside information. What masquerades as inside information is usually no more than a stock tip, or more appropriately, a stock rumor. Small investors love tips and rumors, mistaking them for true inside information.

This is a true story. A friend overheard an employee of a large Denver company in the cafeteria during lunch. He heard that the company was going to be bought out. The stock was sure to go for double what it was presently selling for. He wanted to pass this inside information on to a few close investor friends. The buy-out proposal was true. Some investors did buy that stock immediately. When they had filled their orders on the stock, another item appeared on the local radio news. The purchaser had escaped from a mental hospital that day and was in no position financially or mentally to follow through on the buy-out offer.

This type of inside information usually produces a loss instead of a gain. The important thing to remember is that people with true inside information do not broadcast it. First, it's illegal; and second, it will lose its value if widely known. Generally, it's safe to assume that any hot tips you hear are no more than gossip and rumors—or even worse, deliberately misleading information being distributed by people who have already taken an opposite position on the stock. Tempting though it is to act on such tips, you'll be happier sticking to sound investment principles.

How often in the trading room where I operate have I heard that the president has been shot? Or that someone like Armand Hammer slipped in the bathtub? Later it was revealed that it was his wife who slipped, not he. How these news items can affect the stock market, or one company, is amazing.

Everybody has a good story; brokers, financial advisers, bankers and your friends all have detailed and plausible-sounding explanations for their recommendations. If it didn't make sense to them, they wouldn't be interested in it in the first place. So how does the average investor go about deciding which story to believe? Where and when should you invest your money? Which is a winning stock? Which is a loser?

Insider Trading Is Not Always Illegal

Insider trading can be defined as the buying and selling of a company's stock by its officers, directors, major shareholders, or other people with close ties to the company. These people may legally buy or sell stock of the company for which they are working. The reasons for these insider transactions can be many and are not automatically considered illegal.

Insiders very often receive stock options as part of their regular compensation package. Having exercised their options, when they want cash either to exercise more options or for personal reasons, they sell some of their stock. Thus, they tend to be net sellers most of the time. Insider sale of stock is not necessarily a negative indicator of the company's position, so don't "foul" the next time you see that a corporate insider has sold 500,000 shares of his holdings.

But do say "thank you" when you discover the same person later is buying an additional 500,000 shares for hard dollars. This would be a reason to have a closer look at the company. Insiders recognize low-risk value when they see it. It's an event that deserves further scrutiny when they buy at the market price and not at artificial discounts.

Insiders generally buy under the following conditions:

1. When an announced development will improve the company's profits.
2. When the stock price has fallen so far below its intrinsic values that the shares are cheap.
3. When stock buying is part of a regular investment program.

As an outsider you can make money from any of these situations. However, it's possible for insiders to be too optimistic about the company's future, so this method is not foolproof.

Following insiders is much easier today than it was 30 years ago. Under SEC rules, insiders must report detailed information on their trades within certain time limits. This information is made available to the public in the financial press.

With all thy getting, get understanding. Malcolm S. Forbes

Investigate before you invest. Better Business Bureau

APPENDIX II
HOW TO READ THE
FINANCIAL PAGES

You can either be squirrel food or the seed of a mighty tree. Paul Richey

Stock prices are quoted in most daily newspapers, with emphasis on news of local companies. For extensive market information, the financial press (special business publications) such as the *Wall Street Journal, Investor's Business Daily,* and *Barron's* have even more detailed information. Your daily newspaper will not have all the information the financial press has but it can give you the facts you need.

Stock prices will be shown for both listed and unlisted stocks. Listed stocks are those traded on a stock exchange. The major exchanges are the New York Stock Exchange, the American Stock Exchange, and the NASDAQ (National Association of Securities Dealers Automated Quotation System). Those stocks not listed on an exchange are traded OTC (Over the Counter), and many of these trades are reported.

The information on listed stocks shows:

- High and low for the preceding 52 weeks (not the calendar year). During this period, the stock traded for these prices and there was always a buyer and seller.

- The *stock* name is always abbreviated and listed alphabetically. After you start watching a stock regularly, you can find it very quickly.

- The stock *symbol* identifies companies whose stock is publicly traded. The symbols do not appear in all papers.

- *Dividend* shows the actual cash amount paid annually. The dividend is a payment to shareholders of part of the company's profit. Where no figure is shown, the stock does not pay a dividend.

155

Based on today's price for a stock, most dividends don't appear very exciting. But if you had held the stock for some years, the dividend would represent a large return on your original investment. Also, you would have received dividends all through the years and have an appreciated value on your investment.

- *Yield* is the amount of the dividend expressed as a percentage of the current price of the stock.

- *P/E* (price/earnings ratio). If there is none shown, the company did not earn a profit in the preceding 12 months. The P/E ratio will change whenever the stock's price or earnings per share rises or declines.

- *Volume 100s* are sales in 100s. This figure shows how many shares were traded. Volume or stock market activity is important, though investors should not be concerned about whether a particular stock is or is not on the most active list. Stocks can also rise or fall on low volume.

- *High Low Close.* These three columns show the high, low, and closing price for that day. Usually the differences are small.

- *Net change* shows the change in price from the previous day. The prices for which stocks trade are based on one-sixteenth of a dollar. Each sixteenth of a dollar has a value of 0.0625. Each eighth of a dollar has a value of 0.125. A share of stock priced between $20 and $21 could be 20 $1/8$, 20 $1/4$, 20 $3/8$, 20 $1/2$, 20 $5/8$, 20 $3/4$, or 20 $7/8$ or, another way to say it, 20.125, 20.25, 20.375, 20.50, 20.625, 20.75 or 20.875. To make it simple to understand, estimate each $1/16$ as 6 cents, $1/8$ as 12 cents, so $5/8$ is 60 cents.

OTC stocks, which are usually reported in the same manner as the major exchange stocks, are generally smaller companies that do not meet the listing requirements of the larger exchanges. Some larger companies, however, prefer to stay on the OTC because the financial reporting requirements are not as stringent.

Separate tables in the financial pages highlight the 10 most actively traded stocks for the day, showing the volume of shares traded, the closing price, and the net change.

Periodically, newspapers report earnings of companies. These reports are based on the quarterly or annual reports of the company. Earnings reports are usually made available a few weeks after the quarter ends.

APPENDIX III
GUIDE TO THE ECONOMY AND THE STOCK MARKET

ECONOMIC INDICATORS

Business cycle, inflation, liquidity, and interest rates are key indicators that will help you understand economic influences on the stock market.

1. The business cycle describes the expansion or contraction of the economy as a whole and has an important influence on the earnings trends of most companies. The business cycle affects profitability and cash flow, which is a key element of corporate dividend policy and an element in the movement of the inflation rate up or down. Thus the business cycle affects return on investment.

2. Inflation is the hidden tax we all pay because of the dwindling purchasing power of our dollar. Inflation, which causes prices to rise, is generally caused by excessive government spending. Inflation should be monitored because it's tied to the business cycle. Inflation has a direct impact on investing. A rising or falling inflation rate affects the shift of cash among stocks, bonds, or other alternative investments.

Economists have studied the relationship between the inflation rate and the stock market over long periods of time. They have discovered that the two tend to seek a norm of 20, obtained by adding the market average P/E ratio to the current inflation rate.

Here are some examples from recent decades:

- In the early 1960s, a period of low inflation rates, the market boomed and P/Es rose to 20.
- In the 1970s, when inflation hit 13 percent, the market headed down and P/Es fell to 7.
- During 1997, with an inflation rate of 2 percent, the market hit new highs and P/Es reached 27.

When the combination of P/E and the rate of inflation exceeds or is less than the norm of 20, it suggests that there will be an adjustment in the market.

3. Monetary policy. Federal Reserve Board decisions about money supply liquidity can make the economy grow faster or put a brake on economic growth.

There are several indicators of money supply, but the most useful one for investors to follow is M2—the amount of money in savings and checking accounts. Liquidity and the stock market tend to move together. When M2 grows faster than the economy, stock prices rise. As an example, in the early 1980s when the economy declined sharply, the money supply surged, and the stock market took off.

In 1991, with the economy growing at its slowest rate in 35 years, a mere 1 percent on an inflation-adjusted basis, M2 was growing at 2 percent, or twice the economic growth rate. What happened to the stock market? It made all-time new highs.

4. Interest rates and the stock market move in opposite directions. When interest rates decline, the market does well. When interest rates rise, stock prices tend to decline. Though you can always find winning stocks, it's dangerous to move against the relationship of stocks and interest rates.

How do investors know in which direction interest rates are headed? Look at the Federal Reserve Board (the Fed); it monitors and directs interest rates in the American economy. Look for major turning points in Fed policy, such as a continued discount or interest rate cuts. Don't concern yourself about minor mid-course adjustments.

In 1997, inflation and the market P/E added up to approximately 30. That means that any rise in inflation could have sent the market downward. A further drop in the inflation rate could give the market an upward bias.

Liquidity is positive. If the money supply growth rate accelerates, stocks could move even higher. Still, if it declines, investors should expect stocks to drop from their current lofty levels.

The Fed has moved several times to bring interest rates down. The stock market attracts more investors due to the current interest rate. Investors have to keep informed by reading the newspapers and staying tuned to the direction of any major Fed policy change.

Common stocks would move higher if the economy stays robust and stock earnings increase. If this continues and the economy is indeed improving, higher earnings could be ahead, and the surge in stock prices that we saw in 1992 could be maintained.

Certainly, market cycles are different. The country is entering a new era with the prospects of a peaceful relationship with the Soviet Union and free trade with Canada and Mexico. The extent to which this will affect the stock market remains to be seen. There are signs that inflation could remain low, which would position the U.S. economy for another period of economic expansion.

Being informed of the four major market indicators, and knowing what they mean, will help you as you invest through the years ahead. But the unknowns (politics, etc.) are enough to keep things very interesting.

YOUR WALLET—A LEADING ECONOMIC INDICATOR

What lies behind us and what lies before us are not as important as what lies within us. Anonymous

Consumers account for two-thirds of America's spending and it's easy to understand why we're worth listening to. With our plans to spend, to save, and to invest, we're shaping economic trends. This will help us forecast changes in the economy.

In the late 1970s, the Consumer Expectations Index documented a drop in consumer confidence that reached a low point in mid-1979. Six months later, the economy was in a recession. Consumer expectations and confidence forecast the long expansion nine months before the official onset of the economic recovery in December 1982.

Why is consumer confidence so reliable a forecast? Because you rely on personal experience. You're the one who has to balance your checkbook every month. You see what is happening to interest rates when your CDs mature. It's not the prime rate you care about; it's the mortgage rate. It's not the national unemployment figures that affect you; it's whether you lose your job or anyone in your immediate family or neighborhood loses his. These are much more important to consumers than the figures reported in the newspapers.

In 1997, it was clear that consumers were increasingly concerned about the economy. Today it seems that we're hoping that the fine state of the economy will continue, and we're expecting improvement in the year ahead. But we don't anticipate a roaring recovery. We're paying off debts, saving a little more, and spending a little less.

If you decide to put some of those savings into the stock market as a basis for selling options, reading the financial pages will keep you aware of unusual opportunities for profit. You can buy or sell against the emotions of the crowd.

The hardest thing about making money last is making it first. Anonymous

There are more things in life to worry about than just money . . . how to get hold of it, for example. Anonymous

GLOSSARY

A

Account. A customer of a broker or brokerage firm.

Account number. A number assigned to each investor.

Account statement. A review of all transactions showing the investor's positions during a given period of time. Such statements must be issued quarterly but are also sent at the beginning of each month to active accounts.

Active. A stock or option in which there is heavy volume. Active securities are easier to trade, and the spread between the bid and ask is smaller.

Active market. A heavy volume of trading on an exchange or in the volume of an individual security. An active market helps institutional money managers wishing to buy or sell large positions.

Adjusted option. An option created as the result of a stock split or a stock dividend. For example, after a 2-for-1 stock split, the adjusted option will represent 200 shares.

All-or-none order. An order requiring that it be carried out completely or that none of it be done. This is especially useful for option orders where the volume is small.

American Stock Exchange. (AMEX/ASE) An exchange in New York City where securities of primarily small- to medium-size companies are traded.

American-style option. An option that can be exercised at any time before the expiration date.

Ask. The price at which an option or stock is offered for sale.

Assign. The Options Clearing Corporation notifies a broker-dealer that an option written by one of its customers has been exercised and the terms of settlement must be met. Assignments are normally made in a random manner when the holder of the option contracts asks for delivery of the stock specified in the contract.

Assignment. The receipt of an exercise notice by an option writer (seller) that obligates the writer to sell the underlying security at the specified strike price.

At-the-money. An option that has a strike price equal to the market price of the underlying security.

Averaging down. Buying more shares of the same security at lower prices in order to reduce the average cost.

B

Balance sheet. A financial report that lists the assets, debts, and owners' equity in a company as of a given date. Balance sheets do not list items at their current value, they may exaggerate or understate the real value of certain assets and liabilities. Assets must equal liabilities and equity.

Bankruptcy. The condition of a business when its debts exceed its assets; a state of legal insolvency.

Bear. An investor who believes a stock or the markets in general will follow a broad downward direction. An investors can be a bear on a particular security but not on the general market.

Bear market. A drawn-out period of falling prices in stocks or other assets or in the securities market as a whole, brought on by the expectation of declining economic activity.

Bid. The price a buyer is willing to pay for an option or a stock (bid is to buy, ask is to sell).

Block. A stock trade must total 10,000 shares or more to qualify as a block.

Blue chip. Common stock of a very high quality company having a long history of profit growth and dividend payments.

Borrowing power. The amount of money that may be borrowed in a margin account. The margin limit usually equals 50 percent of the value of all stocks in the account.

Break-even. The price at which a stock or option position can be closed with no profit or loss.

Broker. Person or firm that brings together buyers and sellers for security transactions.

Bull. An investor who believes the price of a particular stock or stocks in general will follow a broad upward trend.

Buy. A bargain-priced security; to purchase a security for money.

Buyback. Purchase of identical securities to cover a short sale or an option call.

Buy/write. The conservative strategy of buying stock and writing covered calls on it at the same time. Example: buying 500 shares of ABC stock at 30, and writing 5 ABC Jan 30 calls.

C

Call. An option contract that gives the owner (the option buyer) the right to buy a specific number of shares at a specified price by a certain date. For the writer (seller) of a call option, the contract is an obligation to sell the specific number of shares if the option is assigned.

Call option. *See* Call.

Call premium. The amount of money that the buyer of a call option has to pay to the seller for the right to purchase stock at a specified price by a certain date.

Capital asset. An asset with an expected life of over one year that is not bought or sold in the normal course of business. Security investments are capital assets.

Capital gain. The difference by which proceeds from the sale of a capital asset exceed its original cost.

Capital gains tax. The tax on profits from the sale of capital assets. The 1997 tax reform law has specified an 18-month holding period after which a capital gain is to be taxed at a maximum of 20 percent for individuals.

Capital loss. The amount by which the cost basis of a capital asset exceeds the proceeds from its sale.

Carry forward. Net capital losses exceeding the annual limit of $3,000 that may be carried to subsequent years so as to offset capital gains or ordinary income. There is no limit to the amount of capital losses that may be used to offset capital gains in any one year, only in the amount of losses in excess of gains that may be used to offset income.

Carrying charge. The interest fee that a broker charges for carrying securities in a margin account.

Cash account. A brokerage account whose transactions are settled on a cash basis, differing from a margin account for which the broker extends credit. Some brokerage investors have both cash and margin accounts. IRA, Keogh and custodial accounts for a minor must be cash accounts.

Cash dividend. A dividend paid in cash to shareholders. The amount is declared by the board of directors and is normally based on profitability; it may exceed net income.

CBOE. The Chicago Board Options Exchange.

Class of options. A term referring to options of the same type, put or call, with the same underlying stock. A class of options having the same expiration date and exercise price is termed a series.

Close. The final price of a security at the end of a trading day. Also called the closing or last price.

Closing buy transaction. The purchase of an option that will eliminate or decrease the size of an open option position previously sold. The obligation to sell stock upon an option assignment will be canceled by the offsetting purchase.

Collateral securities. Shares pledged to a lender, against which margin loans are made. If the value of the brokerage account declines to an unacceptable level, the investor is asked to deliver additional collateral or securities are sold to repay the loan.

Commission. The fee paid to a broker to complete a trade for an investor. Commissions vary among firms.

Common stock. Shares of ownership of a corporation. Owners are entitled to receive dividends and vote for the board of directors. The value of the shares depends on the success of the company.

Confirmation. Written acknowledgment from a broker to a customer of a transaction, giving important details such as date, security, price, commission, and the value of the total order.

Contract. An agreement in call options trading for the writer to sell a specified asset at a set price until a certain date.

Contract size. In options, the amount of the asset to be delivered. One call option contract is for 100 shares.

Contrarian. A strategy of buying and selling securities by going against the crowd. When everyone is buying, contrarians are selling because prices are high, and whenever everyone is selling, contrarians are buying because prices are depressed.

Cover. The purchase of option contracts previously sold.

Covered call income portfolio. The holding of stock on which options can be traded in numerous companies in different industries. The purpose of diversification in such a portfolio is to reduce risk by hedging, acquire capital gains, and receive income from option cash premiums and stock dividends.

Covered call option writing. An investing strategy in which option contracts backed by the underlying shares are written (sold). If the stock price goes up and the option is exercised, the investor has the stock to deliver to the option buyer. Selling covered calls is a conservative strategy that brings in premium income.

Covered option. *See* covered call.

Covered writer. The seller or writer of a call option who owns the stock that may be required for delivery.

D

Day order. An order to buy or sell a stock or option that will expire that day if not completed. All orders are considered day orders unless they are entered as Good-Till-Canceled (GTC).

Debit. An activity in the brokerage account that decreases the cash balance either by purchase of securities or a cash withdrawal.

Debit balance. The margin amount owed in a brokerage account.

Deduction. An expenditure allowed by the Internal Revenue Service that is used to reduce an individual's income-tax-liability.

Deep in/out of the money. A call option that has a strike price below or above the market price of the underlying stock. The option premium for buying a deep in-the-money option is high since the buyer has the right to buy the stock at a strike price below the current market price of the stock. The option premium for buying an out-of-the-money option is low since the buyer has the right to buy the stock at a strike price above the current market price of the stock since the option may never be profitable.

Delist. Removal of a stock or option from an exchange because of failure to meet standards or a merger with another company.

Deliver. The process of meeting the terms of an option contract when an assignment has been executed. The stock has been called; the writer must deliver the stock in one day, and in return receive the strike price. For stock bought or sold, the delivery date is three days after execution.

Derivative. A security whose value is determined in part from the value of another security. An option is a derivative instrument that obtains value from the stock that can be bought with the option.

Discretion. In a brokerage account, discretionary power can be given to an investment advisor or broker to use judgment and transact orders on behalf of the owner of the account.

Diversification. The reduction of risk by purchasing securities of companies in different industries and businesses. (Diversification is just as meaningful to companies as it is to investors.)

Diversified company. A company engaged in different business operations not directly related, or in international trade.

Dividend. A shareholder's allocation of the company's profits distributed. The dividend is paid, usually quarterly, in a fixed amount for each share held as declared by the board of directors.

Dividend coverage. The percentage of earnings paid to shareholders. The higher the percentage, the weaker the dividend coverage, which means the company's management has less flexibility to raise dividends or even to keep them at the same level. Some companies pay out more cash than they earn when earnings decline. Dividend coverage of a stock is a very important item to inspect when investing.

Dividend yield. The annual dividends from a stock divided by its market price per share. The yield gives the return on the current price of the stock, not on the cost basis.

E

Early exercise. A surprise execution before the expiration date. An option holder may decide to exercise early to receive an upcoming dividend.

Earned income. Wages, salaries, commissions, pension, and annuity payments are all taxable as earned income.

Earnings per share. A company's net income divided by each outstanding share of common stock.

Earnings-price (EP) ratio. The net income divided by the current share price, also known as earnings yield. A low EP ratio is used in comparing the appeal of stock growth in earnings; it is the opposite of the price-earnings ratio.

Equity. In a brokerage margin account, the market value of the securities over the debit balance. Equity also refers to ownership in stocks of a corporation.

Equity option. An option for which the underlying asset is stock.

Ex-dividend. A stock no longer carrying the right to the next dividend because settlement occurs after the record date. A stock that has gone ex-dividend is marked with an x in stock transaction tables and in newspaper listings.

Ex-dividend date. The day before an investor must have purchased the stock in order to receive the next dividend. This gives the treasurers of corporations time to mail the checks to shareholders of record by the dividend payment date. This date is referred to as the ex-date and also applies to other situations, such as splits and distributions.

Execution. Carrying out a stock or option trade.

Exercise. The event that occurs when an option owner executes the rights paid for and buys the underlying stock at the strike price.

Exercise—American. Can be exercised at any time on or before the expiration date.

Exercise price. The dollar price at which the owner of an option can buy stock from the seller of an option.

Expiration. The last day on which an option can be exercised. If it is not, the option expires worthless.

Expiration cycle. The three-month cycle of expiration dates. Contracts can be written for one of three cycles: January, February, and March. Options are traded in three-, six-, and nine-month contracts. Only three of the four months in the set are traded at once. Only when the January contract cycle expires will the October contracts start trading.

Cycle	*Expiration months*
January	January-April-July-October
February	February-May-August-November
March	March-June-September-December

Currently, active equity options expire on a hybrid cycle with a total of four option series: this month and next month and the next two months from the above cycle to which that class of options have been assigned. In January options will be trading for January-February-April and July. When January expires, the months outstanding will be February-March-April and July.

Expiration date. The date on which an option right ends.

Expiration month. The month in which an option expires.

F

Fill. When a customer's order to buy or sell securities is executed. When less than the amount of the order is bought or sold, it is a partial fill.

Fungibility. Interchangeable. All options in an option series are *fungible*. All fungible assets such as commodities, options, and securities are interchangeable. For example, shares of ABC Corp. left in custody at a brokerage firm are freely mixed with other shares of ABC Corp. Likewise, stock options are freely interchangeable among investors, just as wheat stored in a grain elevator is not specifically identified as to its ownership.

Fungible. *See* fungibility.

G

Gain. The amount received on a sale of stock that exceeds the amount spent to acquire the stock.

Good-till-canceled (GTC) order. A limit order that remains in effect until the stock or option is executed or canceled.

Greater fool theory. A theory that whatever price an investor pays for a stock or option, another investor with less sense will be willing to buy it at a higher price later. The greater fool theory is most prevalent at the height of a bull market when speculation is at a high.

H

Hedge, hedging. An investment strategy used to offset risk. A shareholder anxious about a declining stock price can hedge by selling a call option.

Highs. Prices for stocks that have reached new peaks for the current 52-week period.

Holder. The owner of a stock or option contract.

Holder of record. The owner of stock as recorded in the company's books as of a certain record date. Dividends and stock splits always specify payment to holders of record as of that date.

House maintenance requirement. The minimum equity that must be kept in a customer's margin account as fixed by the brokerage holding the account. House maintenance sets levels of equity that must be kept to avoid a call for added equity or having the account sold out.

I

Inflation. A rise in prices for goods and services. Inflation is unfavorable to security prices, chiefly because it forces interest rates higher. Two primary U.S. inflation indicators are the Producer Price Index and the Consumer Price Index, which track prices paid by producers and by consumers.

Inside information. A firm's plans that have not been made public. The officers of a company know in advance of favorable or adverse facts that would impact the price of their shares. The Securities and Exchange Commission has rules that an insider is strictly prohibited from trading on the basis of such information.

In-the-money option. An option contract whose underlying stock price is above the strike price of the call option.

Intrinsic value. The in-the-money portion of an option's price. The gap between the exercise price or strike price of an option and the market price of the stock. Options at-the-money or out-of-the-money have no intrinsic value. *See also* in-the-money option.

Investment adviser. A professional offering investment advice as his business; he/she must be registered with the Securities and Exchange Commission.

Investment club. A group of investors who meet regularly and pool their money to invest in securities. Most investment clubs are formed as partnerships. Since investment clubs do not have to file tax returns, all realized capital gains, dividends, and losses are passed through for tax reporting by the individual partners.

Investment income. All income from capital gains on stocks, option premiums, interest, dividends, etc. Interest on margin accounts may be used to offset investment income without limitation.

Investor relations department. Staff of a company who are responsible for filling requests for reports and information from shareholders, prospective shareholders, professional investors, brokers, and the financial media.

L

Last-in, first out. A method for naming the order in which stocks are sold. With last-in, first-out, the items bought most recently are sold first. During a period of inflation the first-in, last-out method tends to result in low cost-basis with high taxable profits.

Last sale. The most recent purchase in a particular security. This should not be confused with the final transaction in a trading day; this is known at the closing sale.

Last trading day. The final day on which an option is traded. Currently, this is the third Friday of the expiration month.

LEAPS (Long-term Equity Anticipation Securities). Call options with an expiration as long as 39 months. Today, equity LEAPS have two series with a January expiration. In November 1997, LEAPS were available with expirations of January 1999 and January 2000.

Leverage. The means for magnifying return or value without increasing investment. Buying stocks or options on margin is leveraging, offering the chance of high returns for a low investment.

Limit order. An order to buy or sell stock or options at an exact price or better.

Limit price. The exact price in a limit order. In buying, the limit price denotes the highest price an investor will pay. In selling, it denotes the lowest price an investor will accept.

Liquidity. The feature of a stock or option where there is enough volume and units outstanding to allow transactions to be filled easily.

Listed option. An option that a national exchange has authorized for trading with standardized terms.

Listed security. A stock or option a national exchange has authorized for trading with standardized terms.

Long term. A holding period of 18 months or longer, according to the Taxpayer Relief Act of 1997; applied in calculating the capital gains tax.

Lows. Prices for stocks that have reached new lows for the current 52-week period.

M

Maintenance call. A call for additional money or securities when the market value of the margin account equity falls below a specified minimum. If the account is not brought up to the required minimum, some of the securities in the account will be sold to bring the account into compliance.

Margin account. A brokerage account that allows investors to buy securities with money borrowed from the broker. Interest is charged on the borrowed funds only for the period of time the loan is used.

Margin agreement. The document that lists the rules of a margin account and that permits a customer to pledge securities in the account as loan collateral.

Margin call. A demand that a customer deposit enough money or securities to bring a margin account up to the minimum maintenance requirements.

Margin/margin requirement. The minimum equity amount a customer is required to deposit with a broker when borrowing to buy securities. Using margin refers to borrowing on securities already in the account or being brought in with the loan being applied to the purchase price. At present the minimum equity requirement is 50 percent of the purchase price, in cash or eligible securities.

Market order. An order to immediately buy or sell a stock or option at the best price available. See limit order for a better price.

Market quote. A current quotation of the bid and ask prices for a stock or option. See last price for a better price.

Most active list. Stocks and options with the most trades on a given day. Unusual volume can be cause for large price movements up or down, depending on the value (good or bad) of the news.

N

Naked option. An option that is sold without the seller owning the under-lying stock. A very high risk position; also known as an uncovered option.

NASD. National Association of Securities Dealers. Trading is done on a national network of computers, instead of an exchange with floor trading.

NASDAQ. National Association of Securities Dealers Automated Quotation system distributes stock quotes from the NASD.

Neutral. A stock, option, or the market in general position that is neither bearish or bullish.

New listing. A stock or option that has recently been added to an exchange. Option listing is decided by the option exchanges, not by the companies that would like to have their names included.

NYSE. The New York Stock Exchange, the largest and oldest stock exchange, also known as the Big Board.

O

OCC. *See* Options Clearing Corporation.

Odd lot. A securities transaction involving less than 100 shares. A round lot is 100 shares.

Offer. Another term for the ask or ask price. Ask is to sell and bid is to buy.

Offset. The closing of an option position by purchasing, for a short posi-tion, or selling for a long position, an identical number of contracts hav-ing the same terms as one already transacted. The offset must be fungi-ble.

Open. An order to buy or sell securities that has not been filled, such as a limit order.

Open interest. The total of option contracts open for an underlying stock. The open interest is reported in newspaper option pages. A large open interest means more activity and liquidity for the option contract.

Opening. The start of a trading day; the first price at which a stock or option trades for the day.

Opening transaction. An option order that starts a new investment position or that increases the size of an existing position. An opening option transaction adds to that option's open interest.

Option. A contract that gives the holder, depending on the type of option held, the right to buy or sell an asset at a set strike price until a specified expiration date in the future. An option to buy an underlying stock is a call and an option to sell an underlying stock is a put. For this right, the call option buyer pays the call option seller, called the writer, a premium fee, which is lost if the buyer does not exercise the option. Buyers of options take on much risk, so they must be correct on the timing and value of the underlying stock to be successful. The option contract obligates the writer to meet the terms of sale if the right is exercised by the holder.

Option agreement. A form filled out by the investor when opening an option account with a brokerage firm. The purpose of the agreement is to assure the broker before trading can begin that the customer qualifies and agrees to the rules and regulations of option trading, and has received an Options Clearing Corporation prospectus.

Option period. The life of an option contract, from the time of selling by a writer to the expiration date.

Option premium. The amount of money paid by an option buyer to the option seller for the right to buy (a call) or sell (a put) the underlying stock at a set price for a specified time. Option writing premiums is a source of additional income.

Optionable stock. A stock on which listed options are traded. Today there are 1,500 optionable stocks and more are being added regularly.

Options Clearing Corporation (OCC). The corporation that handles equity options transactions between buyers and sellers. It is owned by the exchanges. The OCC issues all option contracts and assures that the responsibility of both parties to a trade is fulfilled.

Option series. All option contracts on a stock of the same class, whether puts or calls, that also have the same exercise price and maturity month. An example, all XYZ January 40 calls are a series.

Option writer. The seller of a call option in an opening transaction. The option writer receives a cash premium and incurs an obligation to sell the underlying security at a set price until a specified date, the expiration date.

Ordinary income. Income from wages, dividends, interest and business, as distinguished from capital gains from the sale of assets; income that does not qualify for special tax treatment.

Out-of-the-money. An option whose strike price is higher than the current market price of the underlying security. The option has no intrinsic value; all of its value is time value.

Overvalued. A stock that is trading at a price higher than seems rational given its earnings prospects and the price-earnings ratio. It is difficult to estimate whether or not a security is overvalued.

P

Paper profit or loss. Paper profits and losses are estimated by comparing the current prices of all stocks against all open option contracts in a portfolio to the prices at which those assets were originally bought or sold (the cost basis). These profits or losses are realized when the securities are sold, a taxable event.

Passive income or loss. Endeavors in which a taxpayer does not substantially participate, such as dividends and interest income, as compared to income from wages or an active trade or business.

Payment date. The date on which a stock dividend will be paid.

Payout ratio. The percentage of net income that is paid to shareholders in dividends. Fast-growing corporations reinvest most of their earnings back into the business and usually do not pay dividends. Firms with a high payout ratio have little left for investment in their company to finance future earnings growth.

Portfolio. The holding of stock in numerous companies in different industries. Diversification in a portfolio reduces risk to the investor, allowing him to acquire capital gains and receive dividend income.

Portfolio dressing. Also known in the mutual fund business as window dressing. Addition and removal of securities by an institutional investor before a financial reporting period, usually done quarterly, in order to make the portfolio record appear more satisfactory to investors. This involves the sale of all the big losers and the addition of huge winners to communicate the image that the portfolio manager is extremely talented.

Portfolio manager. An adviser accountable for the investments in a portfolio for an individual, mutual fund, pension plan, bank trust department or insurance company. For a fee, the portfolio manager has the fiduciary obligation to manage the assets prudently and appropriately, selecting stocks, bonds, real estate, or other assets to deliver the best opportunities for profit at any specific time. The author is a portfolio manager.

Position. The combined total status of an investor's stock ownership and open option contracts.

Power of attorney. A document that designates another person to act on behalf of the person signing it. A limited power of attorney only allows transactions within an existing account, but not transfers of the securities to another account. Also known as a discretionary account.

Premium. The price at which an option trades based on time value plus intrinsic value. The amount of the premium is affected by the time to expiration, strike price, and price of the underlying stock. Also known as the option premium.

Premium income. The cash income received when an investor sells a call option, and the option expires worthless or for less than the full premium amount received.

Premium or price quotations. Premium quotes are stated in dollars; a bid or ask of 1 equals $100. Price quotes; the minimum for options trading below 3 is $^1/_{16}$, $6.25, and for all other series $^1/_8$, $12.50, per option contract.

Price. The dollar amount at which a stock or option trades.

Price-earnings (P/E) ratio. The price of a stock divided by its earnings per share. Frequently, a high P/E ratio is an indicator that investors expect the company's earnings to grow. Low P/E stocks tend to be in low-growth or out-of-favor businesses, but this is where the true bargains can be found. No P/E means no earnings.

Put-call ratio. The ratio of trading volume in put options compared with trading volume in call options. This one of the best indicators for stock selection. Option players are not stupid, they are gamblers; so you can see and use what they are doing. A high ratio of puts to calls suggests that the underlying stock is going to go down. A high ratio of calls shows that the underlying stock is expected to go up.

Put option. An option contract that allows the holder the right to sell stock at a specific price for a fixed period of time. This is a mirror image of the covered call option.

Q

Quotation. The bid and ask prices for a stock or an option. Bid is to buy and ask is to sell a security.

Quote. The last price at which a stock or an option traded.

R

Realized gain. The extent to which net proceeds from the sale of a stock or an option exceed its cost. When gains are realized, they become taxable.

Realized loss. The extent to which net proceeds from the sale of a stock or an option are below its cost. When losses are realized, they become taxable.

Reverse Stock Split. A reverse stock split is initiated by companies wishing to raise the price of their outstanding shares, usually to meet the requirements of the exchange or to attract investors who avoid low-priced stocks.

Risk. The uncertainty of profits from an investment, the possibility of losing or not gaining value. The larger the uncertainty of the security price, the greater the risk. Investments with greater implied risk must deliver higher yields.

Round lot. The standard unit of trading—100 shares for stock. For trading lesser amounts, see odd lot.

S

Sale. A sale is "filled," completed, when a buyer and a seller have agreed on a price for the stock or option.

Securities and Exchange Commission (SEC). The agency of the federal government that administers U.S. securities laws and monitors the securities industry.

Security. A stock that shows a holding in a company, or rights to assets such as those represented by an option.

Series of options. Option contracts on the same class, either puts or calls, having the same strike price and expiration month. For example, all XYZ April 30 calls would constitute a series.

Settlement. The transfer of a stock or option for the seller or cash from the buyer in order to complete a transaction.

Settlement day. The time allowed for transfer after execution, currently three days for stocks and one day for options.

Share. A single unit of asset ownership in a company. This ownership is shown by a stock certificate, which designates the company and the shareowner.

Short option position. An investment situation in which the investor has written an option, with the obligation remaining open. The short option position can be closed by buying stock to cover the position with a closing purchase transaction.

Short term. Stock held for 18 months or less, used to distinguish short-term gain or loss from long-term gain or loss to meet tax requirements.

Short-term gain or loss. For tax purposes, the profit or loss realized from the sale of securities held for 18 months or less. Short-term gains are taxable at ordinary income rates.

Speculation. Taking above-average risks to attain above average profits, usually during a comparatively short period of time. Speculation embraces the purchase of something today on the expectation of a higher selling price in the future rather than on its actual value. Buying options is speculation (selling them is not).

Speculator. Anyone eager to take large risks, giving up some safety of principal, to bring in possible large gains. For example, buying short-term options on volatile stocks in hopes of making a large profit in a short time is a speculation.

Spread. The difference between the bid and ask prices for a stock or an option. A large spread often means inactive trading of the security.

Spread order. An option order to trade consisting of two parts to limit risk, such as a buy to close and a sell to open simultaneously.

Stock market. The name that refers to the trading of securities through the various exchanges and the over-the-counter market.

Stock option. The right to buy or sell a stock at a specified price for a specific time. Options offer an opportunity to hedge stock positions, to speculate in stocks with relatively little investment, and to benefit from changes in the market value of the options contracts themselves through an assortment of option strategies.

Stock split. The grant to stockholders of additional shares based on a percentage of current holdings. If, for example, you owned 100 shares of a stock selling at $20 per share, on a 2-for-1 split you would end up with

200 shares valued at $10 per share. A split does not create value. The pie is cut into smaller pieces and though you have more pieces, their value equals what you had previously. The NYSE has found companies that have a stock split usually increase their dividends more than twice as often as "non-splitters."

To identify stock-split candidates, look for one or more of the following factors:

1. a sharply higher market price;
2. a history of stock splits or large stock dividends; or
3. a limited number of shareholders.

Until the new shares are traded and the old ones dropped, the stock is traded "when issued" (wi) and the shares are outstanding though not yet issued. The split shares on the financial page are shown by an "s" after the stock name.

Strike, strike price, exercise. The price at which the owner of a call option can purchase the underlying stock.

Strike price intervals. Prices at which options can be sold. Equity options have $2.50 strike price intervals when the underlying stock is selling for less than $25. When the underlying stock is selling from $25 to $200, the options have $5 intervals. The intervals are $10 when the underlying stock is selling above $200. LEAPS start with one-at-the-money, one in-the-money, and one out-of-the-money strike price.

T

Tax avoidance. Decreasing tax liability by legal means, as in using losses and itemizing deductions.

Tax basis/cost basis. The cost at which a stock or an option was traded, plus brokerage commissions.

Tax loss carry-forward. A tax benefit in which losses incurred in one year can be used to lower tax liability in subsequent years. Individuals may carry over capital losses to offset capital gains and some ordinary income until the losses are used up.

Tax selling. The selling of stocks or options, most often at year-end, to achieve losses in a portfolio that can be used to offset capital gains and thereby lower the tax liability.

Taxable income. All income that is exposed to taxation, calculated after all adjustments and deductions.

Thinly traded. A stock or an option with few bids to buy and few offers to sell, i.e., a very low volume. Investors tend to exclude these securities from their portfolios because their prices are less consistent than those of securities with greater liquidity. Thinly traded options are difficult to transact (to get in or out of a position). Institutional portfolio managers usually ignore such stocks because large orders would have to be done for less than the bid price and for more than the ask price to get fills.

Tick. A price movement, up or down, in a stock or an option.

Time decay. The term used to show how the price of an option wastes away over time.

Time value. That part of the total option premium that represents the time remaining on an option contract before expiration. The premium is constructed of this time value and any intrinsic value of the option.

Total return. On an investment, the consolidation of option premiums, dividends, any capital gain, and tax benefits, stated on an annual basis.

Total volume. The number of shares or option contracts sold in a stock on a given day.

Transaction costs. Expenses for buying or selling stocks or options. These include brokerage commissions, fees for exercise or assignment, exchange fees, SEC fees, and margin interest. The transaction costs to investors who trade can vary greatly depending on the brokerage firm used.

U

Uncovered option. Also known as naked option. An option contract for which the seller does not hold the underlying stock.

Underlying stock. The stock that must be delivered if a call option is exercised.

Undervalued. A stock or an option that trades at a price lower than it rationally should. Whether or not a security is undervalued is an individual consideration. The securities may be cheapened because the business is not well known, the industry is out of favor, or company income is uncertain. Fundamentalists seek to find companies that are undervalued before their stocks become fully or overvalued.

Unit of trading. The least amount of a stock necessary for normal trading purposes, usually 100 shares, a round lot. For option trading, the unit of trade is 100 shares of the underlying stock, unless adjusted for stock splits, stock dividends, or spin-offs.

Unrealized gain (paper gain). The rise in market value of a stock or an option that is still being held, compared with its acquisition cost. Unrealized gains are normally not taxable.

Unrealized loss (paper loss). The reduction in market value of a stock or an option that is still being held, compared with its acquisition cost. Unrealized losses are not recognized for tax purposes.

Unwelcome assignment/surprise assignment. Assignment of an option contract when the writer does not yet choose to perform the terms of the option agreement. Unwelcome assignment is a possibility when calls are trading at or above the strike price and an ex-dividend date is near. Unwelcome assignment can be evaded by buying an offsetting contract and closing the position.

V

Value investing. The preference to buy and sell stocks using the fundamental basis of the value of the company's assets.

Volatile. Subject to huge price changes in either direction. Speculators mostly choose volatile securities only when they buy and sell on short-term price changes.

Volatility. The tendency of a stock or a market to rise and fall swiftly in price within a short time period.

Volume. The total number of shares of stocks and option contracts traded in a given time period. A quick rise in volume may reveal major news has just been announced or is imminent that indicates a future sharp surge or drop in prices.

W

Wasting asset. An asset such as an option in which the expected life is being used up. The risks of a healthy move in the underlying stock lessen as the contract approaches expiration, thus cheapening the value of the option.

When issued (wi). *See* Stock split.

Window dressing. Alteration of a portfolio or financial statement fashioned with dress up activity to create the image of capable administration; e.g., a mutual fund manager may sell losing stocks in his portfolio and replace them with stock positions that have gained in value just before a reporting date.

Write. To sell call option contracts to collect premium income. The writer is the seller. While the option is outstanding, the writer of a call option is obligated by the terms of the contract to sell stock at an agreed strike price by an agreed time.

BIBLIOGRAPHY

There is a wealth of information written about stocks and options. The following list contains the books that I have found the most timeless and useful.

GENERAL INVESTING

Cohen, Jerome B., and Edward D. Zinbarg. *Investment Analysis and Portfolio Management*. Homewood, IL: Dow Jones-Irwin, 1967.

Engel, Louis, and Brendan Boyd. *How to Buy Stocks*. 7th ed. New York: Bantam Books, 1983.

Fisher, Kenneth L. *Super Stocks*. Homewood, IL: Dow Jones-Irwin, 1984.

Goldenberg, Susan. *Trading Inside the World's Leading Stock Exchanges*. San Diego: Harcourt Brace Janovich, 1986.

Graham, Benjamin. *The Intelligent Investor*. 4th ed. New York: Harper & Row Publishers, 1973.

Graham, Benjamin, David L. Dodd, and Sidney Cottle. *Security Analysis*. 4th ed. New York: McGraw Hill Book Co., 1962.

Grant, James. Bernard Baruch: *The Adventures of a Wall Street Legend*. Somerset, NJ: John Wiley & Sons, 1997.

Lefevre, Edwin. *Reminiscences Of A Stock Operator*. Larchmont, NY: American Research Council, 1923.

Little, Jeffrey B., and Lucien Rhodes. *Understanding Wall Street*. 2nd ed. Blue Ridge Summit, PA: Tab Books Inc., 1987.

O'Shaughnessy, James. *What Works On Wall Street*. New York: McGraw-Hill, 1996.

Scott, David L. *Wall Street Words*. Boston, MA: Houghton Mifflin Company, 1988.

Weinstein, Stan. *Secrets for Profiting in Bull and Bear Markets.* Homewood, IL: Dow Jones-Irwin, 1988.

OPTIONS

Angell, George. *Sure Thing Options Trading*. Garden City, NY: Doubleday, 1983.

Ansbacher, Max G. *The New Options Market*. 2nd ed. New York: Walker & Co., 1979.

Gastineau, Gary L. *The Options Manual*. 3rd ed. New York: McGraw-Hill, 1988.

Smith, Courtney. *Option Strategies*. New York: John Wiley & Sons, 1987.

The Options Institute. *Options: Essential Concepts and Trading Strategies*. Homewood, IL: Business One-Irwin, Dow Jones, 1990.

Trester, Kenneth R. *The Compleat Option Player*. 3rd ed. Huntington Beach, CA: Investrek Publishing, 1983.

Walker, Joseph A. *How the Options Markets Work.* New York: New York Institute Of Finance, 1991.

Yates, James W. *The Options Strategy Spectrum.* Homewood, IL: Dow Jones-Irwin, 1987.

Index